PE POWER PRACTICE EXAMS

FOURTH EDITION

JOHN A. CAMARA, PE

Report Errors for This Book

PPI is grateful to every reader who notifies us of a possible error. Your feedback allows us to improve the quality and accuracy of our products. Report errata at **ppi2pass.com**.

PE POWER PRACTICE EXAMS
Fourth Edition

Current release of this edition: 1

Release History

date	edition number	revision number	update
Mar 2021	4	1	New edition.

Printed in the United States of America.

PPI
ppi2pass.com

ISBN: 978-1-59126-788-1

Table of Contents

Preface and Acknowledgments

I wrote *PE Power Practice Exams* to help prepare you for the Principles and Practice of Engineering (PE) Electrical and Computer: Power exam, which is administered by the National Council of Examiners for Engineering and Surveying (NCEES). The two practice exams in this book provide an opportunity for comprehensive exam preparation: the incorrect answers can enlighten; the correct answers are thoroughly explained; and since the problems are based on a range of power engineering topics, you will recognize areas where further study and preparation are needed. With diligent study, you will find the preparation you seek within these pages. At the very least, the questions will lead you to an intelligent search for information. I hope this book serves you well, and that you enjoy the adventure of learning.

I have reviewed all the problems in this book against the NCEES PE Electrical and Computer: Power exam specifications. All of the content was reviewed and, if necessary, revised to comply with the exam-adopted code, the *National Electrical Code* (NEC), 2017 edition. (Where the 2020 NEC differs, the differences are noted for reference.)

Should you find an error in this book, know that it is mine, and that I regret it. Beyond that, I hope two things happen. First, please let me know about the error by using the error reporting form on the PPI website, found at **ppi2pass.com**. Second, I hope you learn something from the error—I know I will! I appreciate suggestions for improvement, additional questions, and recommendations for expansion so that new editions or similar texts will better meet the needs of future examinees.

Thanks go to the very professional and dedicated team at PPI, including

Editorial: Bilal Baqai, Indira Prabhu Kumar, Scott Marley, Scott Rutherford, Grace Wong, Michael Wordelman

Art and Interior Design: Tom Bergstrom

Production: Kimberly Burton-Weisman, Nikki Capra-McCaffrey, Kelly Gunther, Richard Iriye, Damon Larson, Sean Woznicki

Project Management: Beth Christmas

Content and Product: Bonnie Conner, Nicole Evans, Meghan Finley, Anna Howland, Jeri Jump, Megan Synnestvedt, Amanda Werts

Publishing Systems: Sam Webster

And, finally, to Becky Camara, who makes life grand in so many ways.

John A. Camara, PE

Codes and Standards

The information used to write and update this book is based on the exam specifications current at the time of publication. However, just as state and local agencies do not always adopt codes, standards, and regulations as soon as they are issued, the PE exam is not always based on the most current codes. It is likely that the codes that are most current, the codes that you use in practice, and the codes that are the basis of your exam are all different. However, differences between code editions typically minimally affect the technical accuracy of this book, and the methodology presented remains valid. For more information about the variety of codes related to electrical engineering, refer to the following organizations and their websites.

- American National Standards Institute (ansi.org)
- Electronic Components Industry Association (ecianow.org)
- Federal Communications Commission (fcc.gov)
- Institute of Electrical and Electronics Engineers (ieee.org)
- International Organization for Standardization (iso.org)
- International Society of Automation (isa.org)
- National Electrical Manufacturers Association (nema.org)
- National Fire Protection Association (nfpa.org)

The PPI website (**ppi2pass.com**) provides the dates and editions of the codes, standards, and regulations on which NCEES has announced the PE exams are based. It is your responsibility to find out which codes are relevant to your exam.

Electronic versions of the following codes and standards will be provided in their entirety on exam day.

- ANSI C2-2017: *2017 National Electrical Safety Code* (NESC)
- NFPA 30B-2015: *Code for the Manufacture and Storage of Aerosol Products*
- NFPA 70-2017: *National Electrical Code* (NEC)
- NFPA 70E-2018: *Standard for Electrical Safety in the Workplace*
- NFPA 497-2017: *Recommended Practice for the Classification of Flammable Liquids, Gases, or Vapors and of Hazardous (Classified) Locations for Electrical Installations in Chemical Process Areas*
- NFPA 499-2017: *Recommended Practice for the Classification of Combustible Dusts and of Hazardous (Classified) Locations for Electrical Installations in Chemical Process Areas*

These are also listed with bibliographic information in the following section.

CODES AND STANDARDS USED IN THIS BOOK

47 CFR 73: *Code of Federal Regulations*, "Title 47—Telecommunication, Part 73—Radio Broadcast Services," 2018. Office of the Federal Register National Archives and Records Administration, Washington, DC.

IEEE/ASTM SI 10: *American National Standard for Metric Practice*, 2016. ASTM International, West Conshohocken, PA.

IEEE Std. 141 (IEEE Red Book): *IEEE Recommended Practice for Electric Power Distribution for Industrial Plants*, 1993. The Institute of Electrical and Electronics Engineers, Inc., New York, NY.

IEEE Std. 142 (IEEE Green Book): *IEEE Recommended Practice for Grounding of Industrial and Commercial Power Systems*, 2007.

IEEE Std. 241 (IEEE Gray Book): *IEEE Recommended Practice for Electrical Power Systems in Commercial Buildings*, 1990.

IEEE Std. 242 (IEEE Buff Book): *IEEE Recommended Practice for Protection and Coordination of Industrial and Commercial Power Systems*, 2001.

IEEE Std. 399 (IEEE Brown Book): *IEEE Recommended Practice for Industrial and Commercial Power Systems Analysis*, 1997.

IEEE Std. 446 (IEEE Orange Book): *IEEE Recommended Practice for Emergency and Standby Power Systems for Industrial and Commercial Applications*, 1995.

IEEE Std. 493 (IEEE Gold Book): *IEEE Recommended Practice for the Design of Reliable Industrial and Commercial Power Systems*, 2007.

IEEE Std. 551 (IEEE Violet Book): *IEEE Recommended Practice for Calculating Short-Circuit Currents in Industrial and Commercial Power Systems*, 2006.

IEEE Std. 602 (IEEE White Book): *IEEE Recommended Practice for Electric Systems in Health Care Facilities*, 2007.

IEEE Std. 739 (IEEE Bronze Book): *IEEE Recommended Practice for Energy Management in Industrial and Commercial Facilities*, 1995.

IEEE Std. 902 (IEEE Yellow Book): *IEEE Guide for Maintenance, Operation, and Safety of Industrial and Commercial Power Systems*, 1998.

IEEE Std. 1015 (IEEE Blue Book): *IEEE Recommended Practice for Applying Low-Voltage Circuit Breakers Used in Industrial and Commercial Power Systems*, 2006.

IEEE Std. 1100 (IEEE Emerald Book): *IEEE Recommended Practice for Powering and Grounding Electronic Equipment*, 2005.

NEC (NFPA 70): *National Electrical Code*, 2017. National Fire Protection Association, Quincy, MA.

NESC (ANSI C2): *2017 National Electrical Safety Code*, 2017. The Institute of Electrical and Electronics Engineers, Inc., New York, NY.

NFPA 30: *Flammable and Combustible Liquids Code*, 2018. National Fire Protection Association, Quincy, MA.

NFPA 30B: *Code for the Manufacture and Storage of Aerosol Products*, 2015.

NFPA 70E: *Standard for Electrical Safety in the Workplace*, 2018.

NFPA 497: *Recommended Practice for the Classification of Flammable Liquids, Gases, or Vapors and of Hazardous (Classified) Locations for Electrical Installations in Chemical Process Areas*, 2017.

NFPA 499: *Recommended Practice for the Classification of Combustible Dusts and of Hazardous (Classified) Locations for Electrical Installations in Chemical Process Areas*, 2017.

REFERENCES

Anthony, Michael A. *NEC Answers*. New York, NY: McGraw-Hill. (*National Electrical Code* example applications textbook.)

Bronzino, Joseph D. *The Biomedical Engineering Handbook*. Boca Raton, FL: CRC Press. (Electrical and electronics handbook.)

Chemical Rubber Company. *CRC Standard Mathematical Tables and Formulae*. Boca Raton, FL: CRC Press. (General engineering reference.)

Croft, Terrell, and Wilford I. Summers. *American Electricians' Handbook*. New York, NY: McGraw-Hill. (Power handbook.)

Earley, Mark W., et al. *National Electrical Code Handbook*, 2017 ed. Quincy, MA: National Fire Protection Association. (Power handbook.)

Fink, Donald G., and H. Wayne Beaty. *Standard Handbook for Electrical Engineers*. New York, NY: McGraw-Hill. (Power and electrical and electronics handbook.)

Grainger, John J., and William D. Stevenson, Jr. *Power System Analysis*. New York, NY: McGraw-Hill. (Power textbook.)

Horowitz, Stanley H., and Arun G. Phadke. *Power System Relaying*. Chichester, West Sussex: John Wiley & Sons, Ltd. (Power protection textbook.)

Huray, Paul G. *Maxwell's Equations*. Hoboken, NJ: John Wiley & Sons, Inc. (Power and electrical and electronics textbook.)

Jaeger, Richard C., and Travis Blalock. *Microelectronic Circuit Design*. New York, NY: McGraw-Hill Education. (Electronic fundamentals textbook.)

Lee, William C.Y. *Wireless and Cellular Telecommunications*. New York, NY: McGraw-Hill. (Electrical and electronics handbook.)

Marne, David J. *National Electrical Safety Code (NESC) 2017 Handbook*. New York, NY: McGraw-Hill Professional. (Power handbook.)

McMillan, Gregory K., and Douglas Considine. *Process/Industrial Instruments and Controls Handbook*. New York, NY: McGraw-Hill Professional. (Power and electrical and electronics handbook.)

Millman, Jacob, and Arvin Grabel. *Microelectronics*. New York, NY: McGraw-Hill. (Electronic fundamentals textbook.)

Mitra, Sanjit K. *An Introduction to Digital and Analog Integrated Circuits and Applications*. New York, NY: Harper & Row. (Digital circuit fundamentals textbook.)

Parker, Sybil P., ed. *McGraw-Hill Dictionary of Scientific and Technical Terms*. New York, NY: McGraw-Hill. (General engineering reference.)

Plonus, Martin A. *Applied Electromagnetics*. New York, NY: McGraw-Hill. (Electromagnetic theory textbook.)

Rea, Mark S., ed. *The IESNA Lighting Handbook: Reference & Applications*. New York, NY: Illuminating Engineering Society of North America. (Power handbook.)

Shackelford, James F., and William Alexander, eds. *CRC Materials Science and Engineering Handbook*. Boca Raton, FL: CRC Press. (General engineering handbook.)

Van Valkenburg, M.E., and B.K. Kinariwala. *Linear Circuits*. Englewood Cliffs, NJ: Prentice-Hall. (AC/DC fundamentals textbook.)

Wildi, Theodore, and Perry R. McNeill. *Electrical Power Technology*. New York, NY: John Wiley & Sons. (Power theory and application textbook.)

Introduction

ABOUT THIS BOOK

PE Power Practice Exams contains two practice exams designed to match the format and specifications defined by the National Council of Examiners for Engineering and Surveying (NCEES). Each 80-problem practice exam is divided into two four-hour-long sessions. A step-by-step solution is provided for each problem. Each solution presents the solving method needed to arrive at the answer, along with illustrations, explicit calculations, and relevant assumptions and code references.

Solutions presented for each problem may represent only one of several methods for obtaining the correct answer. Alternative problem-solving methods may also produce correct answers.

In the solutions, each equation from the *NCEES Handbook* is given in blue and annotated with the title of the section the equation is found in, also in blue. Whenever data are taken from a figure or table in the *NCEES Handbook*, the title of the figure or table is given in blue. Get to know these titles as you study; they will give you search terms you can use to quickly find the equations and data you need, saving valuable time during the exam.

Equations in red are also essential knowledge for the exam, but they are either not found in the *NCEES Handbook* or not found there in their most basic forms. Some of these are fundamental equations—such as Ohm's law, Kirchhoff's laws, the relationship between torque and power, and the formulas for resistors, inductors, and capacitors in series and in parallel—that you should be sure you know well, as you won't be able to refresh your memory of them during the exam. Some equations are simplifications or common derivations of equations in the *NCEES Handbook*, and you are likely to need to know and apply the common alternative form for the exam. Some are required background knowledge that will help you understand knowledge areas you will be tested on in the exam. All these are marked in red so that you know to study these equations and concepts as well.

ABOUT THE EXAM

The PE Electrical and Computer: Power exam (the Power exam) is made up of 80 problems and is divided into two four-hour sessions. Each session contains 40 multiple-choice problems. Only one of the four options given is correct, and the problems are completely independent of each other. Problems that are related to codes and standards will be based on either (1) an interpretation of a code or standard that is presented in the exam booklet, or (2) a code or standard that a committee of licensed engineers feels minimally competent engineers should know.

The topics and the distribution of problems for the Power exam are as follows. The distribution for both the practice exams and the NCEES exam is approximate.

- **General Power Engineering (21–32 questions)**

  measurement and instrumentation (4–6 questions); applications (7–11 questions); codes and standards (10–15 questions)

- **Circuits (14–21 questions)**

  analysis (8–12 questions); devices and power electronic circuits (6–9 questions)

- **Rotating Machines and Electric Power Devices (14–21 questions)**

  induction and synchronous machines (7–11 questions); electric power devices (7–11 questions)

- **Transmission and Distribution (21–32 questions)**

  power system analysis (10–15 questions); protection (11–17 questions)

For further information and tips on how to prepare for the Power exam, consult PPI's website, **ppi2pass.com**.

HOW TO USE THIS BOOK

Prior to taking the practice exams in this book, assemble your materials as if you were taking the actual exam. Refer to NCEES for a list of the materials that will be provided to you during the exam. These will include the NEC; the NESC; NFPA 70E-2018, *Standard for Electrical Safety in the Workplace*; and the NFPA hazardous area classification standards—all listed in *NCEES Handbook* Sec. 2.3, Codes and Standards. Be sure to confirm that you are studying with the indicated version or edition of each of these codes, as these are not always the most current versions, and obtain copies of the appropriate reference or code books. Be sure to visit **ppi2pass.com/stateboards** to find a link to your state's board of engineering, and check for any state restrictions on materials you are allowed to use during the exam.

The two exams in this book allow you to structure your exam preparation the way it works best for you. You may choose to take one exam as a pretest to assess your knowledge and determine the areas in which you need more review, and then take the other after you have completed additional studying. You may instead choose to take both exams after you have completed your studying. Regardless of how you decide to use this book, when you are ready to begin, set a timer for four hours and take the morning session. Use the calculator and references you have gathered for use on the exam, and mark your answer on the answer sheet. After a one-hour break, turn to the afternoon session, set the timer for another four hours, and complete the afternoon session. When the timer goes off, check your answers and review the solutions to any problems you answered incorrectly or were unable to answer. Compare your problem-solving approaches against those given in the solutions. Once you feel you are sufficiently prepared, set the timer again and take the second exam, marking your answers on the answer sheet. If you feel you need more review after taking both practice exams, check **ppi2pass.com** for the latest in exam preparation materials and resources.

Morning Session Instructions

In accordance with the rules established by your state, you may use any approved battery- or solar-powered, silent calculator to work this examination. However, no blank papers, writing tablets, unbound scratch paper, or loose notes are permitted. Sufficient paper will be provided. The *NCEES PE Electrical Power Reference Handbook* and provided codes are the only references you are allowed to use during this exam.

You are not permitted to share or exchange materials with other examinees.

You will have four hours in which to work this session of the examination. Your score will be determined by the number of questions that you answer correctly. There is a total of 40 questions. All 40 questions must be worked correctly in order to receive full credit on the exam. There are no optional questions. Each question is worth 1 point. The maximum possible score for this section of the examination is 40 points.

Partial credit is not available. No credit will be given for methodology, assumptions, or work written on scratch paper.

Record all of your answers on the Answer Sheet. Mark your answers with a no. 2 pencil. Answers marked in pen may not be graded correctly. Marks must be dark and must completely fill the bubbles. Record only one answer per question. If you mark more than one answer, you will not receive credit for the question. If you change an answer, be sure the old bubble is erased completely; incomplete erasures may be misinterpreted as answers.

If you finish early, check your work and make sure that you have followed all instructions. After checking your answers, you may submit your answers and leave the examination room. Once you leave, you will not be permitted to return to work or change your answers.

When permission has been given by your proctor, you may begin your examination.

Do not work any questions from the Afternoon Session during the first four hours of this exam.

Name: ______________________________________
Last First Middle Initial

Examinee number: ______________

Examination Booklet number: ______________

Principles and Practice of Engineering Examination

Morning Session
Practice Exam 1

Morning Session

1.	A	B	C	D	11.	A	B	C	D	21.	A	B	C	D	31.	A	B	C	D
2.	A	B	C	D	12.	A	B	C	D	22.	A	B	C	D	32.	A	B	C	D
3.	A	B	C	D	13.	A	B	C	D	23.	A	B	C	D	33.	A	B	C	D
4.	A	B	C	D	14.	A	B	C	D	24.	A	B	C	D	34.	A	B	C	D
5.	A	B	C	D	15.	A	B	C	D	25.	A	B	C	D	35.	A	B	C	D
6.	A	B	C	D	16.	A	B	C	D	26.	A	B	C	D	36.	A	B	C	D
7.	A	B	C	D	17.	A	B	C	D	27.	A	B	C	D	37.	A	B	C	D
8.	A	B	C	D	18.	A	B	C	D	28.	A	B	C	D	38.	A	B	C	D
9.	A	B	C	D	19.	A	B	C	D	29.	A	B	C	D	39.	A	B	C	D
10.	A	B	C	D	20.	A	B	C	D	30.	A	B	C	D	40.	A	B	C	D

Exam 1: Morning Session

1. Review the following power transmission one-line diagram.

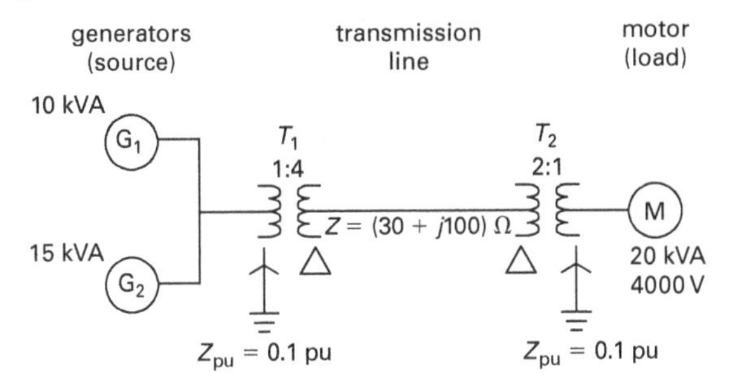

The 10 kVA rating of generator 1 is taken as the base. What is most nearly the per-unit kVA power of the motor?

(A) 0.5 pu

(B) 1.0 pu

(C) 1.5 pu

(D) 2.0 pu

2. An 11,000 V wye-connected generator with a grounded neutral supplies an 11,000 V delta/460 V wye-connected transformer. Per *National Electrical Code* requirements, the transformer wye is grounded. Which of the following surge arrester connections would provide the best protection for the primary side distribution system in the event of a lightning strike?

(A) high-resistance neutral connection on the generator wye

(B) high-resistance ground connection on the transformer wye

(C) phase-to-ground connection on the transformer primary side

(D) phase-to-phase connection on the transformer secondary side

3. A resistor converts electrical energy to heat at a rate of 10.8 kJ/min. The resistor has 900 C/min passing through it. What is most nearly the resistor voltage drop?

(A) 12 V

(B) 24 V

(C) 84 V

(D) 120 V

4. A three-phase step-down transformer rated at 3600 kVA has a per-unit impedance of 7.5% listed on the manufacturer's nameplate. The turns ratio is 25 with a primary voltage rating of 11 kV. Assume the line values are the base values. What is most nearly the actual impedance on the secondary side?

(A) 0.004 Ω

(B) 0.8 Ω

(C) 4 Ω

(D) 900 Ω

5. In the circuit shown, the time, t, is measured in seconds.

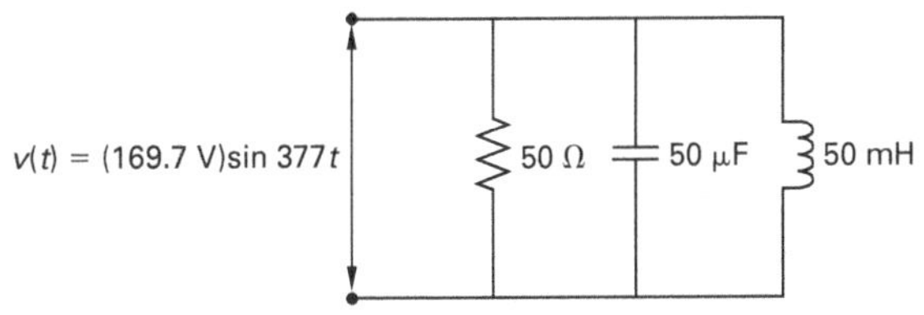

What is most nearly the equivalent impedance of the circuit shown?

(A) $15\ \Omega\angle 0°$

(B) $25\ \Omega\angle 60°$

(C) $35\ \Omega\angle 60°$

(D) $50\ \Omega\angle -45°$

6. The basic lightning impulse insulation level is defined in terms of

(A) the withstand voltage for a specified transient

(B) a lightning impulse withstand voltage

(C) a standard switching impulse

(D) a standard lightning impulse

7. During an emergency, a makeshift circuit is created using two separate power supplies, as shown.

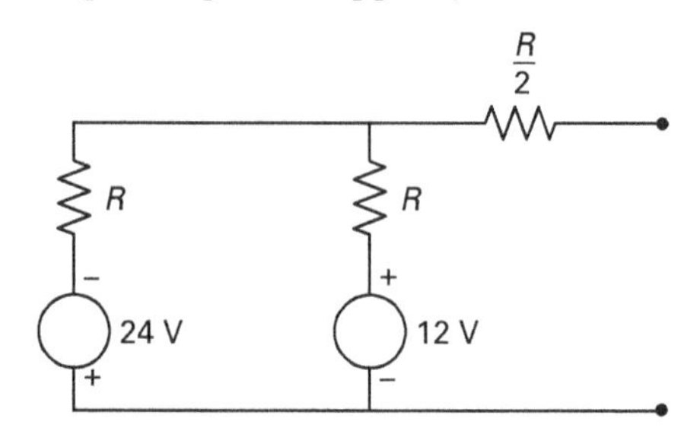

What is the Thevenin equivalent resistance of the indicated circuit?

(A) $R/2$

(B) R

(C) $2R$

(D) R^2

8. An electrical distribution system has the following properties.

three-phase, four-wire
wye connection, neutral ground
480 V line voltage

A complex unbalanced load of $500+j300$ VA is connected across phase A and phase C. What is most nearly the neutral current?

(A) 1.2 A

(B) 2.1 A

(C) 2.4 A

(D) 4.2 A

9. A simplified Thevenin equivalent diagram of a single-phase distribution system is shown. At time $t=0$ s, a fault occurs, inserting 0 Ω in parallel with the load resistance.

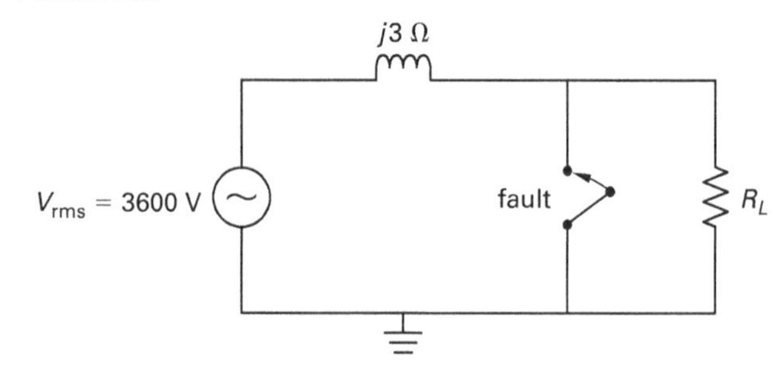

What is most nearly the expected rms current a few cycles into the fault?

(A) 1100 A

(B) 1200 A

(C) 3600 A

(D) indeterminate

10. A 60 Hz, three-phase, 208 V system carries a total load of $200+j150$ VA. What is most nearly the capacitance required to correct the power factor?

(A) 0.001 μF

(B) 3 μF

(C) 9 μF

(D) 500 μF

11. The individual control system for a small tooling machine draws 5 A. How many of the control systems may be cord connected to a single quad receptacle on a four-receptacle circuit rated for 20 A?

(A) 1

(B) 2

(C) 3

(D) 4

12. A 60 Hz, three-phase, 208 V system uses a capacitor to correct the power factor. The capacitor is rated at 440 V and 150 kVAR. What is most nearly the reactive power provided by the capacitor?

(A) 35 kVAR

(B) 57 kVAR

(C) 75 kVAR

(D) 670 kVAR

13. Two vectors, representing an electric and magnetic field in a material, are given by the following.

$$\mathbf{E} = 3\mathbf{i}+7\mathbf{j}+3\mathbf{k} \text{ V/m}$$
$$\mathbf{H} = 2\mathbf{i}-3\mathbf{j}+\mathbf{k} \text{ A/m}$$

Determine a vector orthogonal to both **E** and **H**.

(A) $-2\mathbf{i}+3\mathbf{j}-5\mathbf{k}$

(B) $16\mathbf{i}+9\mathbf{j}-5\mathbf{k}$

(C) $16\mathbf{i}+9\mathbf{j}-23\mathbf{k}$

(D) $16\mathbf{i}+3\mathbf{j}-23\mathbf{k}$

14. The system shown was designed as an ungrounded system, but a resistor was added in the dotted box, allowing a path to ground via the neutral.

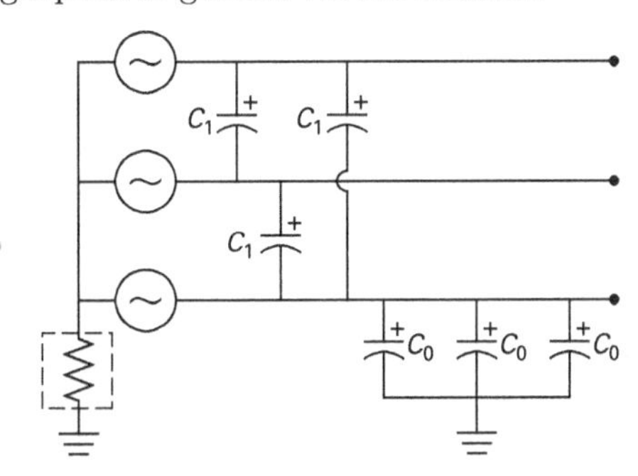

What is the purpose of the resistor?

(A) to allow for the creation of a voltage drop and limit unbalanced current flow

(B) to allow sufficient ground current for detection of a capacitive ground fault

(C) to limit ground current and prevent damage during fault conditions

(D) both (A) and (B)

15. The d'Arsonval meter shown is configured to perform as a DC voltmeter with a sensitivity of 1000 Ω/V and an accuracy of 1%. The coil resistance is 250 Ω.

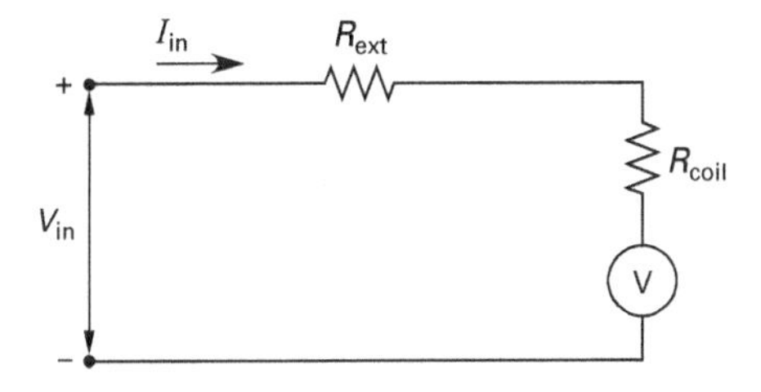

What is most nearly the maximum external resistance value allowable for the meter to read 10 V at full scale with an accuracy of 2.5%?

(A) 9800 Ω

(B) 9850 Ω

(C) 9900 Ω

(D) 9950 Ω

16. A hemispheric lamp is mounted 5 m above the ground. The lamp is rated for 3000 lm. What is most nearly the illumination on the ground directly below the lamp?

(A) 20 lx

(B) 100 lx

(C) 600 lx

(D) 3000 lx

17. A portion of the program coding for a PLC application is shown. The input line is hot.

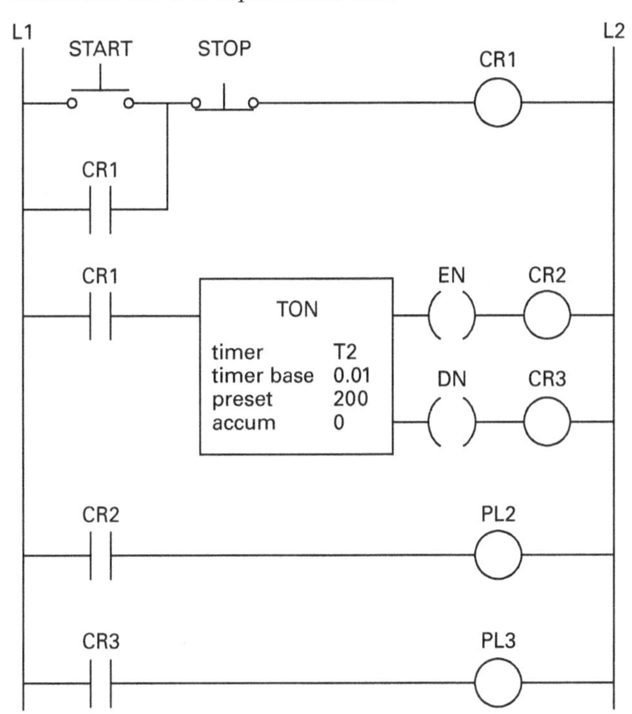

Which conditions must be met for the rule for rung 3 to be true or ON?

(A) CR1 normally open contacts close; timer accum 200

(B) CR1 normally open contacts close; timer DN line ON

(C) CR1 normally open contacts close; EN relay ON; CR2 ON

(D) START momentary switch closed; CR1 ON; EN and CR2 relays ON

18. A 208 V, wye-connected, ungrounded system suffers a ground at phase A. What is most nearly the line voltage at phase B?

(A) 120 V

(B) 150 V

(C) 210 V

(D) 360 V

19. A control system will operate in an environment with an average temperature of 150°F. The copper wire supplying the system is IACS annealed copper with an alpha value of 0.00402°C^{-1}, a reference temperature of 20°C, and a resistance of approximately 1 Ω per 1000 ft. The system is designed to operate at 220 V DC and 30 A. To ensure proper operation, this control system can experience no more than a 3.0 V input drop between the source and the circuit. What is most nearly the maximum length of IACS copper wire used in such a two-wire DC system?

(A) 10 m

(B) 12 m

(C) 13 m

(D) 26 m

20. A 33 kV to 12 kV transformer has a delta-connected primary and a wye-connected secondary. The transformer is rated for 50 MVA with a 3.5% per-unit impedance. The transformer is in a distribution system with multiple transformers. To convert the distribution system to a one-line diagram, the base power is selected as 100 MVA. What is the per-unit impedance on the 100 MVA base?

(A) 2%

(B) 3%

(C) 7%

(D) 9%

21. For the given instrument transformer, which closely approximates an ideal transformer, the supply frequency is 60 Hz.

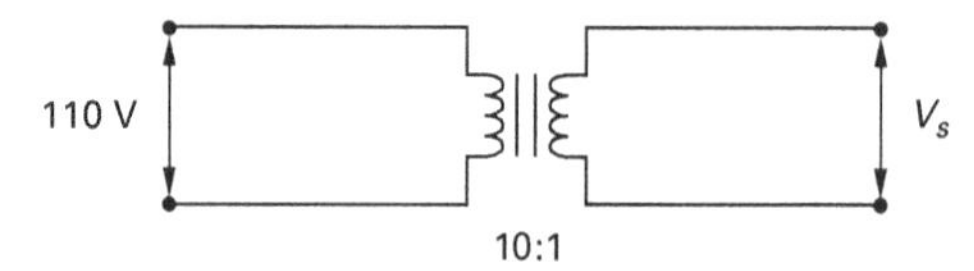

What is most nearly the value of the peak magnetic flux in the secondary?

(A) 6.0 mWb

(B) 11 mWb

(C) 40 mWb

(D) 45 mWb

22. Two three-phase transformers are rated for 480 V secondary voltage. One transformer has a delta connection and the other has a grounded wye connection on the secondary side. Which of the following statements is FALSE?

I. The transformers may be connected in parallel.

II. The delta connection is more fault tolerant.

III. The wye connection ensures voltages remain constant during ground conditions.

(A) I only

(B) I and II

(C) II only

(D) II and III

23. A single-family dwelling unit of 2000 ft^2, excluding unfinished attic and garage space, is planned. What is the minimum number of 20 A branch circuits required?

(A) 3

(B) 4

(C) 5

(D) 6

24. A 373 kW induction motor operating at a speed of 600 rpm has a 0.8 power factor. A 480 V, three-phase, 60 Hz system supplies the motor. The starting torque is 125% of the motor's full-load torque. What is most nearly the starting torque of the motor?

(A) 750 N·m

(B) 1900 N·m

(C) 5900 N·m

(D) 7400 N·m

25. A 50 mV shunt is used to detect the current during a battery discharge. The discharge profile is as follows.

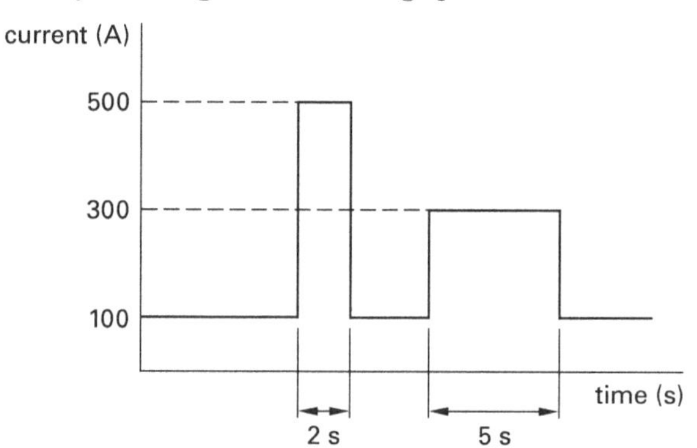

What is the approximate shunt resistance?

(A) 1 $\mu\Omega$

(B) 100 $\mu\Omega$

(C) 300 $\mu\Omega$

(D) 500 $\mu\Omega$

26. Which of the following characteristics of a battery determines the correct charging voltage?

(A) age

(B) chemistry

(C) temperature

(D) both (B) and (C)

27. A single-family dwelling unit of 2000 ft^2, excluding unfinished attic and garage space, is planned. Using the standard calculations, what is most nearly the net load (including general lighting, small appliance, and laundry loads), with any applicable demand factors applied?

(A) 5600 VA

(B) 6000 VA

(C) 7500 VA

(D) 9000 VA

28. A 200 hp induction motor operating at 1150 rpm has a 0.8 power factor and a no-load speed of 1190 rpm and draws 175 A. The system supplying the motor is rated for 480 V, three-phase, and 60 Hz. What is most nearly the motor's speed regulation?

(A) 3.4%

(B) 3.5%

(C) 97%

(D) Speed regulation is not applicable to motors.

29. A single-family dwelling unit has a net load of 4000 VA. A 6 kW dryer and 8 kW range will also be included in the design. The dwelling unit is supplied by a 120/240 V, three-wire, single-phase copper feeder of type THW. According to the *National Electrical Code*, what size feeder wire is required?

(A) AWG 3

(B) AWG 4

(C) AWG 6

(D) AWG 8

30. A synchronous generator is rated for 15 MVA, 11 kV, and 0.8 lagging power factor. At rated conditions, what is most nearly the real power output?

(A) 12 MW

(B) 15 MW

(C) 21 MW

(D) 26 MW

31. Consider the given nameplate data from an auto-start motor-operated appliance.

item	parameter
phase	1
power	¾ hp
voltage	115 V
current	16 A
service factor	1.15

What is the appropriate level for overload protection?

(A) 10 A

(B) 15 A

(C) 20 A

(D) 25 A

32. Consider the following induction motor's speed-versus-torque curve.

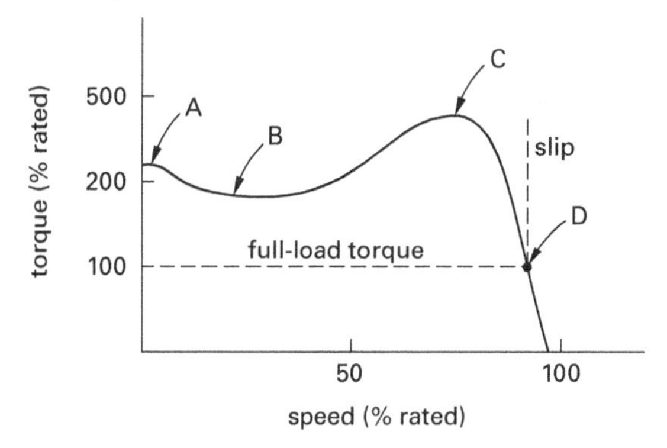

What is point B?

(A) full-load torque

(B) locked-rotor torque

(C) pull-out torque

(D) pull-up torque

33. A proposed branch circuit for a manufacturing plant expansion has the following design parameters and restrictions.

item	parameter/restriction
intermittent load	12 A
continuous load	10 A
THHN conductors	90°C
overcurrent device	60°C terminal device
ambient temperature	50°C maximum
raceway	5 conductors

What is the appropriate conductor size for the branch circuit?

(A) AWG 6

(B) AWG 8

(C) AWG 10

(D) AWG 12

34. The figure shown represents the physical configuration of three phases on a 13.8 kV distribution line.

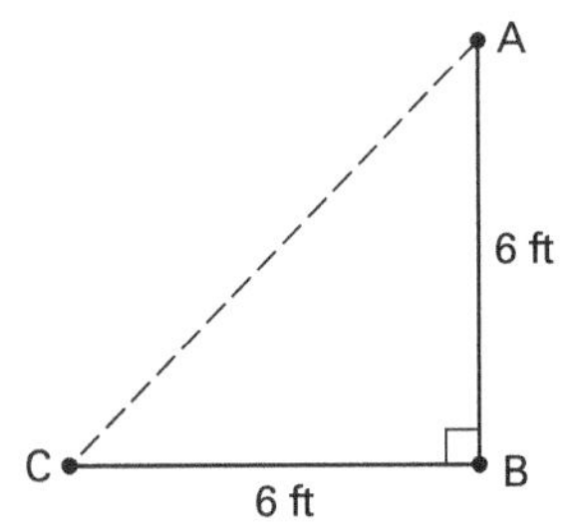

The geometric mean distance is required to determine the correct tables to use in the *National Electrical Safety Code*, since they are tabulated using symmetrical spacing. What is most nearly the geometric mean distance for the configuration shown?

(A) 6 ft

(B) 7 ft

(C) 8 ft

(D) 9 ft

35. A 5 hp, 230 V, three-phase squirrel-cage motor has a nameplate full-load current rating of 14 A. It is a Design A motor with a service factor of 1.00. What is most nearly the minimum conductor ampacity required?

(A) 15.2 A

(B) 16.1 A

(C) 17.5 A

(D) 19.0 A

36. Consider the following partial transmission system.

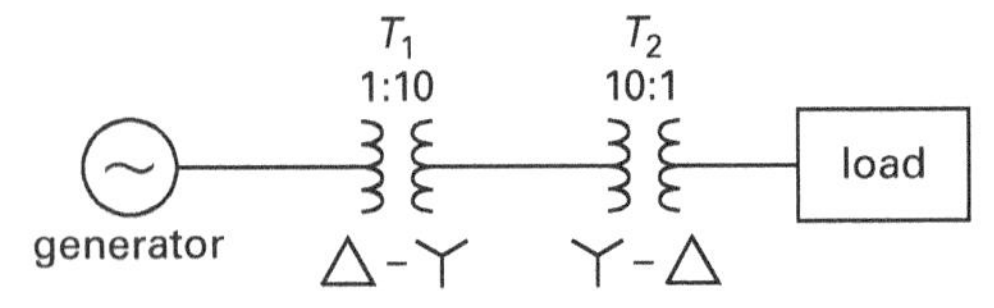

The generator is rated at 100 MVA with an output voltage of 23 kV. Transformer T_1 is rated at 150 MVA. The manufacturer's nameplate lists the impedance as 8%. Transformer T_2 is rated for 100 MVA with an impedance of 11%. Using transformer values as the base, what is most nearly the base impedance on the primary side of T_1?

(A) 1.5 Ω

(B) 3.5 Ω

(C) 5.3 Ω

(D) 8.2 Ω

37. A 5 hp, 230 V, three-phase squirrel-cage motor has a nameplate full-load current rating of 14 A. It is a Design A motor with a service factor of 1.00. What is most nearly the maximum setting of the separate overload device?

(A) 15.2 A

(B) 16.1 A

(C) 17.5 A

(D) 19.0 A

38. Consider the partial transmission shown. The load is 75 MVA, and the power factor is 0.8 lagging.

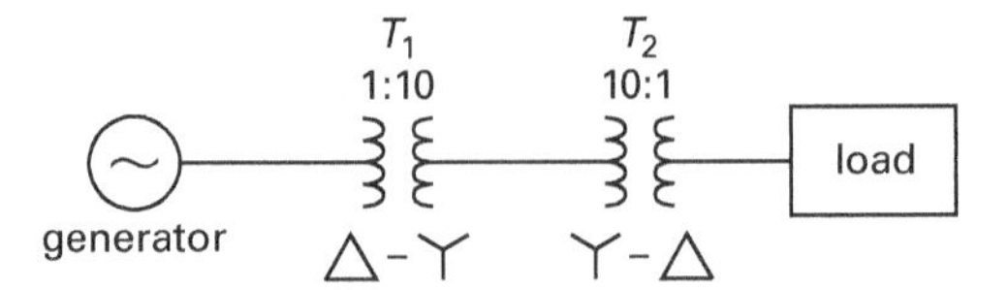

Which of the following statements is FALSE?

(A) The resistance of the generator is ignored in first-order calculations.

(B) Per-unit impedances on T_1 are the same from the primary to the secondary side.

(C) The angle between the voltage and current at the generator is −37°.

(D) The angle between the voltage and current in the transmission line is −67°.

39. A 5 hp, 230 V, three-phase squirrel-cage motor has a nameplate full-load current rating of 14 A. It is a Design A motor with a service factor of 1.00. If a non-time-delayed fuse is used as the short-circuit and ground fault protection, what is most nearly the appropriate fuse rating?

(A) 15 A

(B) 20 A

(C) 45 A

(D) 50 A

40. A three-phase, four-wire system has a single load between phase A and phase C. The line voltage is 11 kV, and the phase voltage is 6.35 kV. The load is 300 kVA, and the power factor is 0.8 lagging. What is most nearly the line current in phase A?

(A) 21.8 A

(B) 27.3 A

(C) 38.6 A

(D) 47.3 A

STOP!

DO NOT CONTINUE!

This concludes the Morning Session of the examination. If you finish early, check your work and make sure that you have followed all instructions. After checking your answers, you may submit your answers and leave the examination room. Once you leave, you will not be permitted to return to work or change your answers.

Afternoon Session Instructions

In accordance with the rules established by your state, you may use any approved battery- or solar-powered, silent calculator to work this examination. However, no blank papers, writing tablets, unbound scratch paper, or loose notes are permitted. Sufficient paper will be provided. The *NCEES PE Electrical Power Reference Handbook* and provided codes are the only references you are allowed to use during this exam.

You are not permitted to share or exchange materials with other examinees.

You will have four hours in which to work this session of the examination. Your score will be determined by the number of questions that you answer correctly. There is a total of 40 questions. All 40 questions must be worked correctly in order to receive full credit on the exam. There are no optional questions. Each question is worth 1 point. The maximum possible score for this section of the examination is 40 points.

Partial credit is not available. No credit will be given for methodology, assumptions, or work written on scratch paper.

Record all of your answers on the Answer Sheet. Mark your answers with a no. 2 pencil. Answers marked in pen may not be graded correctly. Marks must be dark and must completely fill the bubbles. Record only one answer per question. If you mark more than one answer, you will not receive credit for the question. If you change an answer, be sure the old bubble is erased completely; incomplete erasures may be misinterpreted as answers.

If you finish early, check your work and make sure that you have followed all instructions. After checking your answers, you may submit your answers and leave the examination room. Once you leave, you will not be permitted to return to work or change your answers.

When permission has been given by your proctor, you may begin your examination.

Do not work any questions from the Morning Session during the second four hours of this exam.

Name: ______________________________
Last First Middle Initial

Examinee number: ______________

Examination Booklet number: ______________

Principles and Practice of Engineering Examination

Afternoon Session Practice Exam 1

Afternoon Session

41.	(A)	(B)	(C)	(D)
42.	(A)	(B)	(C)	(D)
43.	(A)	(B)	(C)	(D)
44.	(A)	(B)	(C)	(D)
45.	(A)	(B)	(C)	(D)
46.	(A)	(B)	(C)	(D)
47.	(A)	(B)	(C)	(D)
48.	(A)	(B)	(C)	(D)
49.	(A)	(B)	(C)	(D)
50.	(A)	(B)	(C)	(D)
51.	(A)	(B)	(C)	(D)
52.	(A)	(B)	(C)	(D)
53.	(A)	(B)	(C)	(D)
54.	(A)	(B)	(C)	(D)
55.	(A)	(B)	(C)	(D)
56.	(A)	(B)	(C)	(D)
57.	(A)	(B)	(C)	(D)
58.	(A)	(B)	(C)	(D)
59.	(A)	(B)	(C)	(D)
60.	(A)	(B)	(C)	(D)
61.	(A)	(B)	(C)	(D)
62.	(A)	(B)	(C)	(D)
63.	(A)	(B)	(C)	(D)
64.	(A)	(B)	(C)	(D)
65.	(A)	(B)	(C)	(D)
66.	(A)	(B)	(C)	(D)
67.	(A)	(B)	(C)	(D)
68.	(A)	(B)	(C)	(D)
69.	(A)	(B)	(C)	(D)
70.	(A)	(B)	(C)	(D)
71.	(A)	(B)	(C)	(D)
72.	(A)	(B)	(C)	(D)
73.	(A)	(B)	(C)	(D)
74.	(A)	(B)	(C)	(D)
75.	(A)	(B)	(C)	(D)
76.	(A)	(B)	(C)	(D)
77.	(A)	(B)	(C)	(D)
78.	(A)	(B)	(C)	(D)
79.	(A)	(B)	(C)	(D)
80.	(A)	(B)	(C)	(D)

Exam 1: Afternoon Session

41. What is the determinant of the following matrix?

$$\mathbf{A} = \begin{bmatrix} 3 & 4 & 2 \\ 1 & 6 & 5 \\ 4 & 3 & 2 \end{bmatrix}$$

(A) −120

(B) −21

(C) 0

(D) 21

42. A single-phase resistive load draws 5 kW with a terminal voltage of 220 V. The load is supplied by a single-phase source through 180 ft lines, with an impedance of $0.8 + j0.5\ \Omega$ per 1000 ft. What is most nearly the source voltage?

(A) 220 V

(B) 230 V

(C) 240 V

(D) 250 V

43. An engineering project (design A) requires an investment of $200 million and provides a lump sum return of $350 million in 10 yr. An alternative (design B) requires an investment of $60 million and provides a lump sum return of $80 million in 5 yr. Another option (design C) requires an investment of $20 million and provides a lump sum return of $30 million in 20 yr. Assume an interest rate of 5%. Which alternative is economically superior?

(A) design A

(B) design B

(C) design C

(D) all options are economically equivalent

44. A two-pole induction motor operates at 60 Hz. If the slip is 3%, what is most nearly the operating speed?

(A) 110 rpm

(B) 1800 rpm

(C) 3500 rpm

(D) 3600 rpm

45. Two circuits with inductive components are mounted closely together. When $v_1 = (169.7\ \text{V})\sin 377t$ and $v_2 = (33.95\ \text{V})\sin 377t$, mutual inductance becomes a significant factor. (t is measured in seconds.)

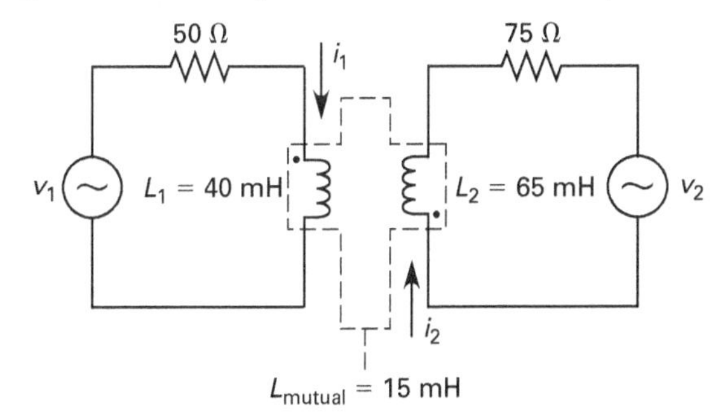

Both voltage sources are de-energized, and an impedance meter is connected in place of v_1. What is most nearly the measured impedance of the circuit?

(A) $50\ \Omega\angle 0°$

(B) $52\ \Omega\angle 17°$

(C) $54\ \Omega\angle 23°$

(D) $80\ \Omega\angle 22°$

46. A 5 hp, 208 V, three-phase induction motor draws 17 A with a power factor of 0.8 from a 220 V system. What is most nearly the reactive power required to correct the power factor to 0.9?

(A) −3890 VAR

(B) −2510 VAR

(C) −1380 VAR

(D) −1320 VAR

47. The primary-to-secondary turns ratio for an instrument transformer supplying a high-impedance electronic circuit is 33:1.

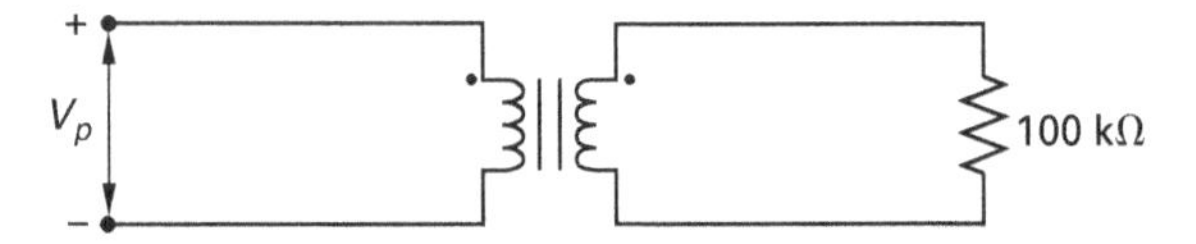

The supply voltage on the primary is 110 V. The transformer and core approximate the ideal. What is most nearly the magnitude of the current on the primary?

(A) 1 μA

(B) 1 mA

(C) 1 A

(D) 10 A

48. A three-phase capacitor bank will be used to correct the power factor from 0.7 to 0.8 lagging. The load is 5000 kW. What is most nearly the reactive power required to make the correction?

(A) 1350 kVAR

(B) 3460 kVAR

(C) 4390 kVAR

(D) 5150 kVAR

49. Assume that an incandescent 100 W lightbulb is on for an average of 6 h/d. The average cost of electricity per kilowatt hour is $0.07. What is most nearly the annual cost of electricity used to operate the lightbulb?

(A) $1.50/yr

(B) $5.50/yr

(C) $10.00/yr

(D) $15.00/yr

50. What is the Thevenin equivalent voltage between terminal A and terminal B in the following 60 Hz circuit?

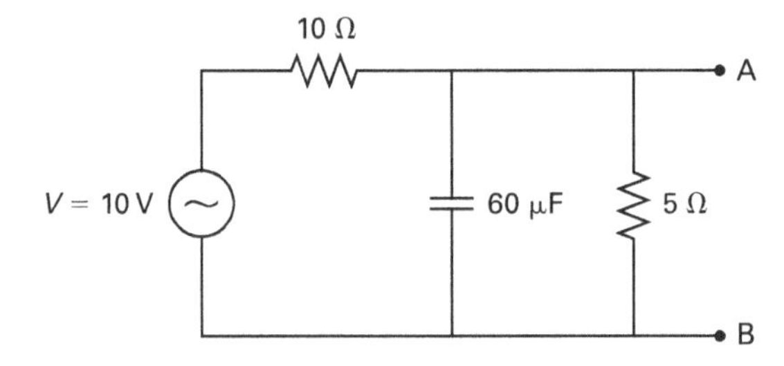

(A) 3.3 V∠−4°

(B) 3.3 V∠−90°

(C) 5.0 V∠−4°

(D) 5.0 V∠−6°

51. Two generators comprise the electrical supply system for a given geographical area. Generator A has a 2% probability of failure. Generator B, which operates independently of generator A, is slightly less reliable, with a 4% probability of failure. *Failure* is defined as a problem of sufficient magnitude to result in removal of the generator from the electric grid. One generator is required to ensure peak power demand is met. What is the likelihood that peak power demand will be met?

(A) 0.0008

(B) 0.0400

(C) 0.9800

(D) 0.9992

52. What type of conductor is intended to carry current under normal conditions, even though the vectorial sum of all system phases is zero potential at the conductor's connection point?

(A) bonding jumper

(B) grounded conductor

(C) grounding electrode

(D) neutral conductor

53. A typical utility source setup and the service entrance panel to a dwelling are shown.

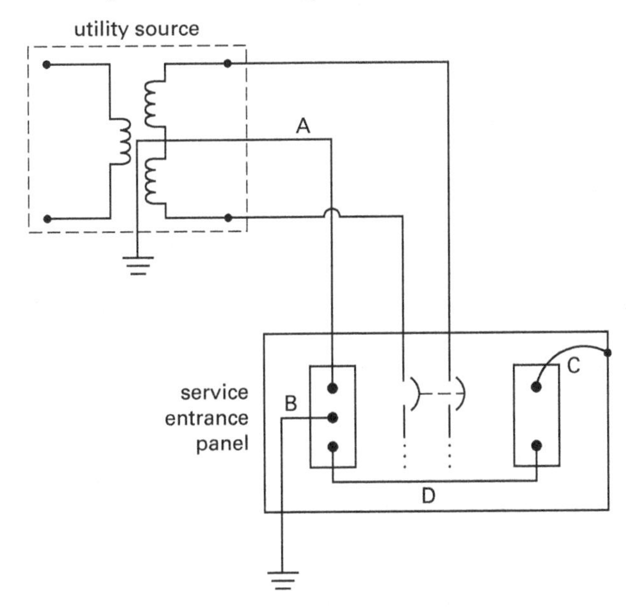

The grounding electrode conductor is defined in NEC Art. 100 as the "conductor used to connect the system grounded conductor or the equipment to a grounding electrode or to a point on the grounding electrode system." Which of the labeled conductors is the grounding electrode conductor?

(A) conductor A

(B) conductor B

(C) conductor C

(D) conductor D

54. The following one-line schematic represents a three-phase transformer. What is most nearly the turns ratio, primary to secondary?

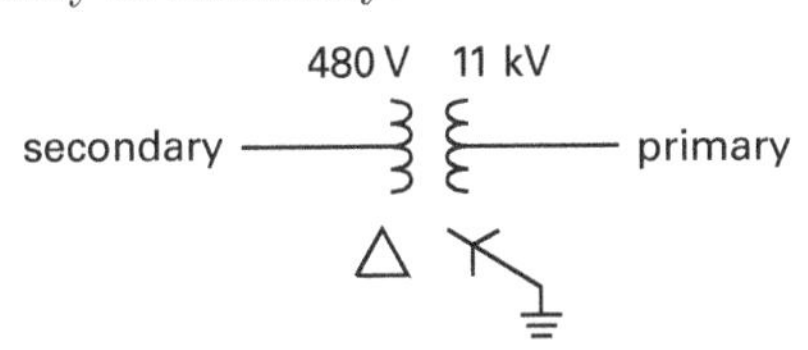

(A) 13

(B) 23

(C) 37

(D) 43

55. The following op amp circuit is used to supply a resistive load. The op amp characteristics are nearly ideal. The feedback resistance, R_f, is 1 MΩ. The input resistance, R_{in}, is 25 Ω. The load resistance, R_L, is 50 Ω.

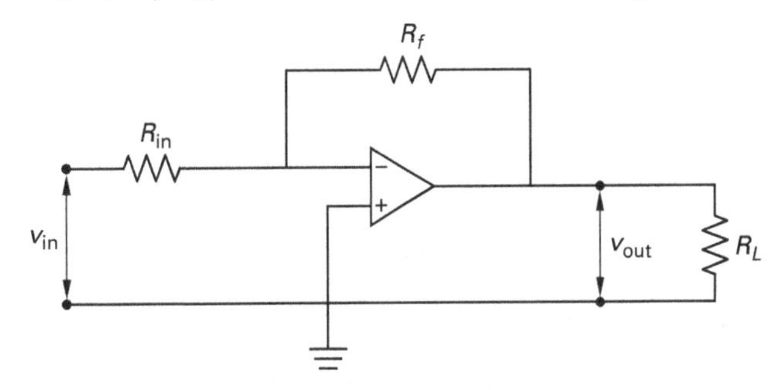

What is most nearly the voltage gain of the circuit?

(A) −40 000

(B) −20 000

(C) 2.0

(D) 13 000

56. An induction motor is rated for 10% slip at full load and has a nameplate speed of 1620 rpm when connected to a 60 Hz source. What is the number of poles?

(A) 2

(B) 4

(C) 6

(D) 8

57. An ideal transformer powers a 3.0 $\Omega\angle 30°$ load, as shown.

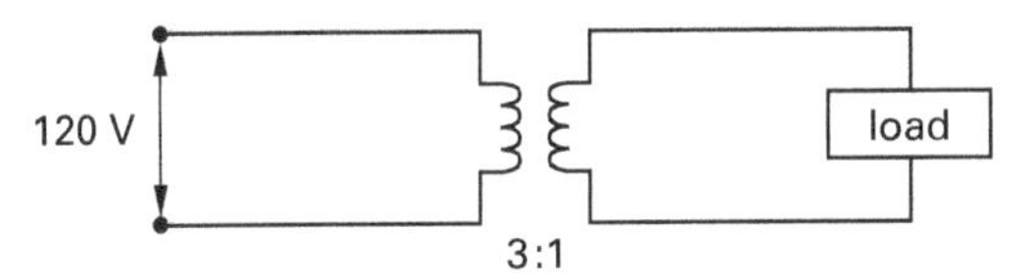

What is most nearly the steady-state impedance seen by the 120 V source?

(A) 9.0 $\Omega\angle -30°$

(B) 9.0 $\Omega\angle 30°$

(C) 27 $\Omega\angle 30°$

(D) 40 $\Omega\angle -150°$

58. According to the NESC, what is the required illumination level for a large, centralized control room of an electric supply station?

(A) 15 lx

(B) 25 lx

(C) 160 lx

(D) 270 lx

59. Which of the following conditions must exist for a ground-fault circuit interrupter (GFCI) to remove power to its associated receptacle?

(A) unbalanced current

(B) unbalanced magnetic fields

(C) induced voltage

(D) all of the above

60. An electrical bus has a full-load voltage of 1.00 pu and a no-load voltage of 1.03 pu. What is the voltage regulation on the bus?

(A) 3.00%

(B) 5.00%

(C) 97.0%

(D) 103%

61. For a transmission line with the following properties, what is most nearly the characteristic impedance?

$$L_l = 500 \ \mu\text{H/mi}$$
$$C_l = 0.100 \ \mu\text{F/mi}$$

(A) $50.0 \times 10^{-12} \ \Omega$

(B) $7.07 \times 10^{-6} \ \Omega$

(C) $1.40 \times 10^{-2} \ \Omega$

(D) $70.7 \ \Omega$

62. In the circuit shown, phase C is open. The currents flowing in phase A and phase B are shown, with phase A as the reference.

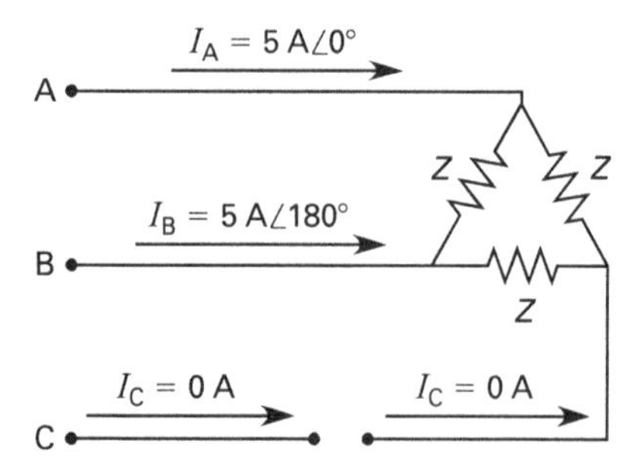

What is most nearly the phase A positive sequence current, I_A?

(A) 3 A∠−30°

(B) 3 A∠30°

(C) 5 A∠30°

(D) 9 A∠−30°

63. An arrester provides a protection level of 38 kV. Let the minimum protective margin allowed by standards be 20%. The equipment to be protected has a withstand voltage of 90 kV. What is most nearly the protective margin?

(A) 20%

(B) 40%

(C) 60%

(D) 140%

64. A current transformer relay that is one part of a differential zone protection scheme is designed to trip at 160 A worth of primary current. The transformer's turns ratio is 400:5. The secondary current is within the burden of the transformer. What relay tap setting is required for proper operation?

(A) 2.0 A

(B) 5.0 A

(C) 80 A

(D) 160 A

65. A section of a wye-connected three-phase transmission line has a phase voltage rating of 15 kV and a power rating of 30 kVA. The per-phase line impedance is 75 Ω. What is the per-unit impedance?

(A) 0.01 pu

(B) 2.5 pu

(C) 3.0 pu

(D) 7.5 pu

66. Certain fundamental assumptions are made in nearly all stability studies. Which of the following is NOT one of those fundamental assumptions?

(A) Faults are limited to load components only.

(B) Generated voltage is unaffected by machine speed variations.

(C) Only synchronous currents and voltages are considered in windings.

(D) Symmetrical components are used in the representation of unbalanced faults.

67. A distribution system is designed as a three-phase, four-wire, neutral-grounded system. The phase-to-phase voltage is 12.8 kV. The complex load power is expected to be $150 + j75$ kVA between phase A and neutral and between phase B and neutral. The expected neutral current due to phase A and B is most nearly

(A) 0.0 A

(B) 13 A

(C) 23 A

(D) 33 A

68. A single-phase distribution system uses two conductors placed 0.3 m apart. Each wire has a radius of 20 mil. The system will be utilized at 10 GHz. What is most nearly the capacitive reactance per unit length?

(A) $4 \times 10^{-12}\ \Omega/\text{m}$

(B) $7 \times 10^{-12}\ \Omega/\text{m}$

(C) $4\ \Omega/\text{m}$

(D) $7\ \Omega/\text{m}$

69. Which of the following statements referring to a SCADA system is FALSE?

(A) The system is a subsystem of the overall energy management system (EMS) of some electric power networks.

(B) The system provides data acquisition, supervisory control, and alarm display.

(C) The latest systems use the open systems interconnection (OSI) protocol.

(D) The system is used for automatic protection of a network; manual control is a separate function.

70. Which of the following statements about triplen harmonics is FALSE?

(A) Triplen harmonics include the 3rd, 9th, and 15th harmonics.

(B) Triplen harmonics cause increased heating in wiring.

(C) Triplen harmonics cancel one another at the wye neutral connection.

(D) Delta connections prevent triplen harmonics from affecting line quantities.

71. Of the following power disturbances, which one represents a harmonics problem?

(A)

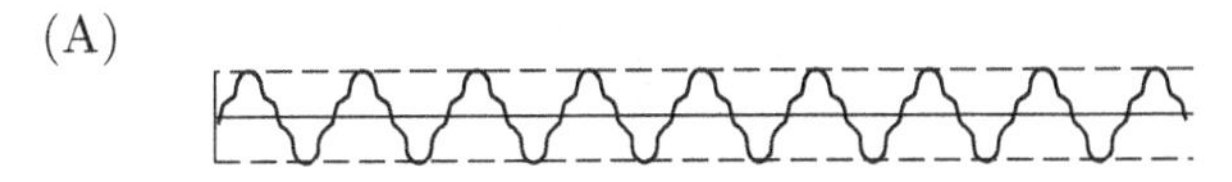

(B)

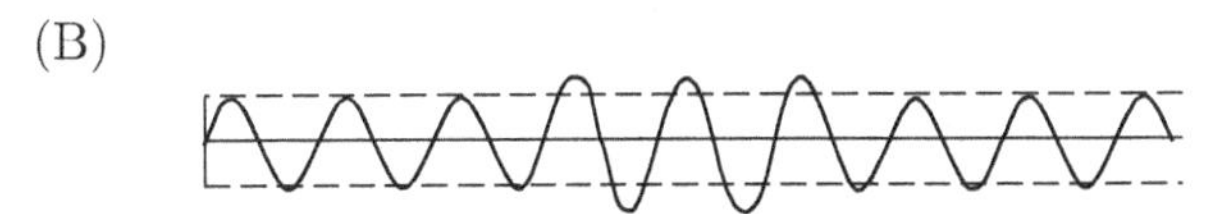

(C)

(D)

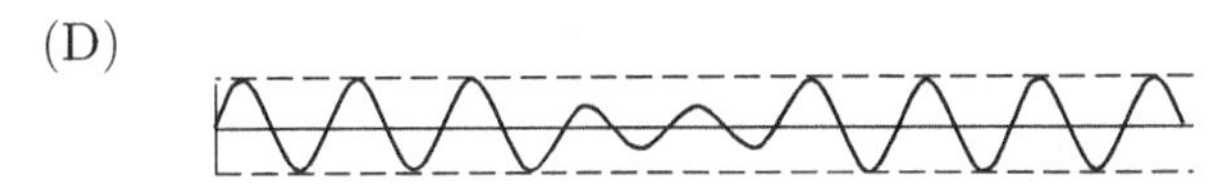

72. A single-phase transformer with a primary rated voltage of 12.2 kV and a secondary voltage of 480 V has a per-unit impedance of 7%. During a short-circuit test, what is most nearly the voltage that will be applied to the high-voltage terminals?

(A) 33 V

(B) 480 V

(C) 850 V

(D) 12,200 V

73. A three-phase wye-connected system with a grounded neutral has a single load connected between phase A and phase C. The line voltage is 220 V. The load is 3.3 kVA at 0.80 lagging power factor. What is most nearly the magnitude of the current in phase A?

(A) 10 A

(B) 15 A

(C) 22 A

(D) 26 A

74. Which of the following is NOT an advantage of lithium-ion batteries?

(A) high energy-to-weight ratio

(B) slow self-discharge rate

(C) no memory effect

(D) minimal cost

75. A given circuit is meant to carry a continuous 10 A load. In addition, three loads designed for a permanent display stand are fastened in place and require 1 A each when operating. What is most nearly the rating of the overcurrent protective device (OCPD) on the branch circuit?

(A) 15 A

(B) 20 A

(C) 25 A

(D) 30 A

76. A 12 V battery with an internal resistance of 0.5 Ω supplies a motor that is approximately a 2 Ω load. Assuming a fully charged battery, what is most nearly the load current?

(A) 5.0 A

(B) 6.0 A

(C) 12 A

(D) 24 A

77. In the following graph of torque versus speed, what type of DC motor does the curve represent?

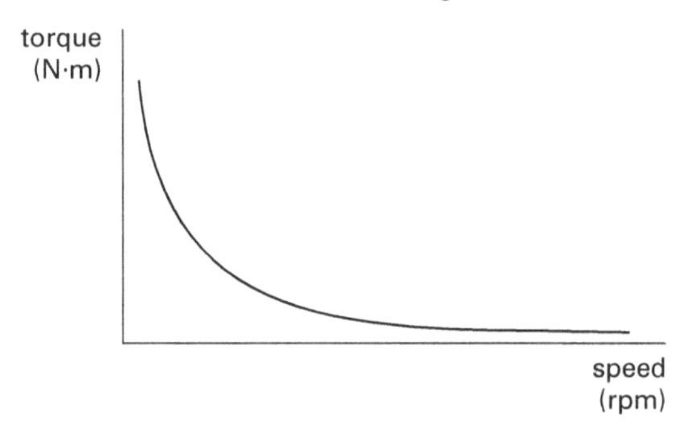

(A) cumulatively compounded

(B) differentially compounded

(C) series wound

(D) shunt wound

78. What is the array utilization of the programmable logic array (PLA) shown?

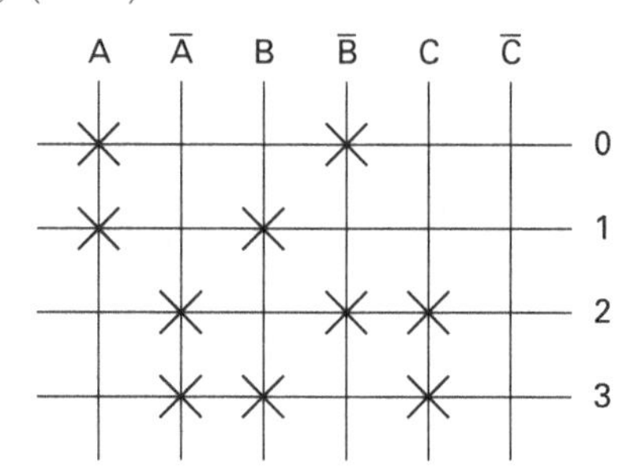

(A) 25%

(B) 33%

(C) 42%

(D) 75%

79. For a four-pole generator that produces a 60 Hz output, what is most nearly the speed of the armature?

(A) 240 rpm

(B) 1200 rpm

(C) 1800 rpm

(D) 3600 rpm

80. Which of the following transformer losses are core losses and are not dependent on the load?

(A) copper losses

(B) eddy current losses

(C) hysteresis losses

(D) both (B) and (C)

STOP!

DO NOT CONTINUE!

This concludes the Afternoon Session of the examination. If you finish early, check your work and make sure that you have followed all instructions. After checking your answers, you may submit your answers and leave the examination room. Once you leave, you will not be permitted to return to work or change your answers.

Morning Session Instructions

In accordance with the rules established by your state, you may use any approved battery- or solar-powered, silent calculator to work this examination. However, no blank papers, writing tablets, unbound scratch paper, or loose notes are permitted. Sufficient paper will be provided. The *NCEES PE Electrical Power Reference Handbook* and provided codes are the only references you are allowed to use during this exam.

You are not permitted to share or exchange materials with other examinees.

You will have four hours in which to work this session of the examination. Your score will be determined by the number of questions that you answer correctly. There is a total of 40 questions. All 40 questions must be worked correctly in order to receive full credit on the exam. There are no optional questions. Each question is worth 1 point. The maximum possible score for this section of the examination is 40 points.

Partial credit is not available. No credit will be given for methodology, assumptions, or work written on scratch paper.

Record all of your answers on the Answer Sheet. Mark your answers with a no. 2 pencil. Answers marked in pen may not be graded correctly. Marks must be dark and must completely fill the bubbles. Record only one answer per question. If you mark more than one answer, you will not receive credit for the question. If you change an answer, be sure the old bubble is erased completely; incomplete erasures may be misinterpreted as answers.

If you finish early, check your work and make sure that you have followed all instructions. After checking your answers, you may submit your answers and leave the examination room. Once you leave, you will not be permitted to return to work or change your answers.

When permission has been given by your proctor, you may begin your examination.

Do not work any questions from the Afternoon Session during the first four hours of this exam.

Name: ____
Last First Middle Initial

Examinee number: ______________

Examination Booklet number: ______________

Principles and Practice of Engineering Examination

Morning Session Practice Exam 2

Morning Session

81. (A) (B) (C) (D)
82. (A) (B) (C) (D)
83. (A) (B) (C) (D)
84. (A) (B) (C) (D)
85. (A) (B) (C) (D)
86. (A) (B) (C) (D)
87. (A) (B) (C) (D)
88. (A) (B) (C) (D)
89. (A) (B) (C) (D)
90. (A) (B) (C) (D)
91. (A) (B) (C) (D)
92. (A) (B) (C) (D)
93. (A) (B) (C) (D)
94. (A) (B) (C) (D)
95. (A) (B) (C) (D)
96. (A) (B) (C) (D)
97. (A) (B) (C) (D)
98. (A) (B) (C) (D)
99. (A) (B) (C) (D)
100. (A) (B) (C) (D)
101. (A) (B) (C) (D)
102. (A) (B) (C) (D)
103. (A) (B) (C) (D)
104. (A) (B) (C) (D)
105. (A) (B) (C) (D)
106. (A) (B) (C) (D)
107. (A) (B) (C) (D)
108. (A) (B) (C) (D)
109. (A) (B) (C) (D)
110. (A) (B) (C) (D)
111. (A) (B) (C) (D)
112. (A) (B) (C) (D)
113. (A) (B) (C) (D)
114. (A) (B) (C) (D)
115. (A) (B) (C) (D)
116. (A) (B) (C) (D)
117. (A) (B) (C) (D)
118. (A) (B) (C) (D)
119. (A) (B) (C) (D)
120. (A) (B) (C) (D)

Exam 2: Morning Session

81. A three-phase delta load has a phase current of $10\ \text{A}\angle 5°$. The line voltage is $120\ \text{V}\angle 30°$. What is most nearly the phase reactance?

(A) 5.1 Ω

(B) 10 Ω

(C) 12 Ω

(D) 15 Ω

82. An electric generator is rated for 20 MVA, 13.8 kV. It is a diesel engine-driven, three-phase, four-wire system. The generator is a salient pole with dampers. The manufacturer's data for reactances is

$$\begin{aligned}\text{subtransient direct reactance, } X_d'' &= 0.20\\ \text{transient direct reactance, } X_d' &= 0.30\\ \text{direct reactance, } X_d &= 1.25\\ \text{quadrature reactance, } X_q &= 0.90\end{aligned}$$

Assume a terminal voltage of 1.0 pu for a transient study conducted at rated power and voltage. The power factor is 1.0. Most nearly, what magnitude of generated voltage should be used for this study?

(A) 1.02 pu

(B) 1.04 pu

(C) 1.35 pu

(D) 1.60 pu

83. A sinusoidal voltage with an rms value of 110 V is applied to a load impedance of $6 + j3\ \Omega$. What is most nearly the real power delivered to the load?

(A) 0.60 kW

(B) 0.80 kW

(C) 1.6 kW

(D) 1.8 kW

84. The ground fault relay is set to operate when it senses 5 V on the secondary. To ensure accuracy, the secondary load is below the burden of the current transformer. What value of neutral-to-ground voltage is required to trip the relay and cause a protective function?

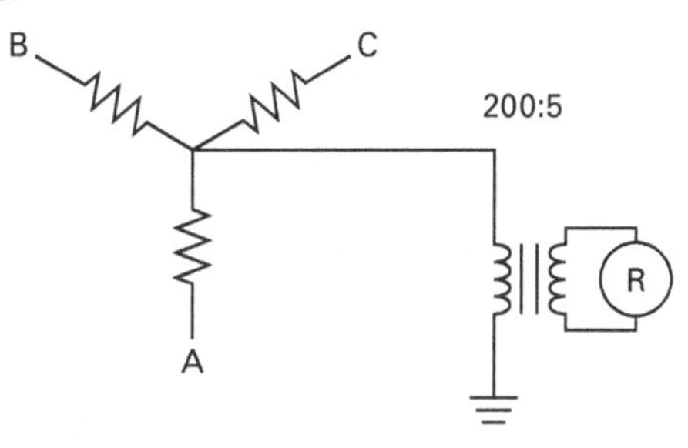

(A) 0.2 kV

(B) 0.4 kV

(C) 0.6 kV

(D) 0.8 kV

85. The units associated with the unlabeled power triangle shown are W, VA, and VAR.

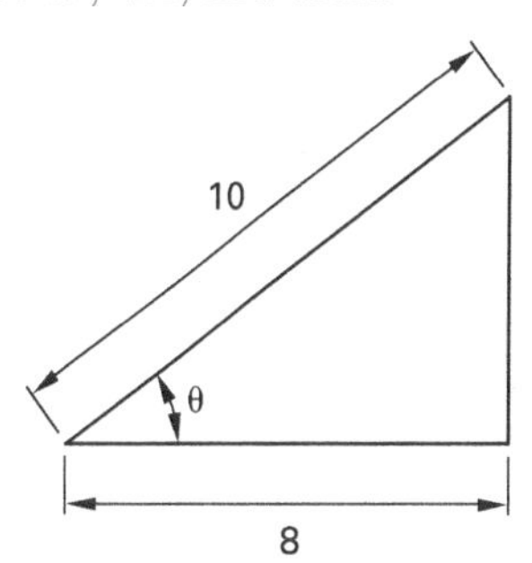

What is most nearly the magnitude of the reactive power?

(A) 1 VAR

(B) 6 VAR

(C) 10 VAR

(D) 13 VAR

86. An overcurrent relay is designed to protect a circuit from a three-phase, 5000 A fault. The current transformer (CT) sensing the fault has a turns ratio of 100:1 and utilizes the 2.5 A tap. The total burden is 1.5 Ω at the expected fault condition. What is the secondary voltage (i.e., the excitation voltage) of the current transformer during the fault condition?

(A) 50 V

(B) 75 V

(C) 110 V

(D) 220 V

87. A transformer, which can be closely modeled as ideal, is shown.

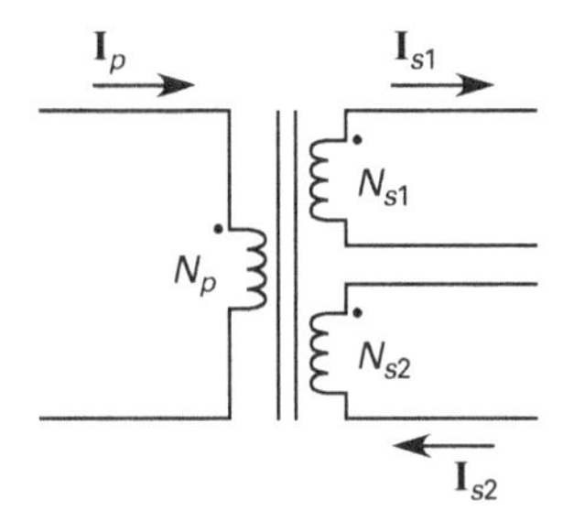

The turns ratio for N_p:N_{s1} and N_p:N_{s2} is 12:1. The currents in the transformer secondary are measured as $\mathbf{I}_{s1} = 5 \text{ A}\angle 30°$ and $\mathbf{I}_{s2} = 6 \text{ A}\angle 10°$. What is most nearly the current in the primary winding?

(A) $0.80 \text{ A}\angle 20°$

(B) $0.90 \text{ A}\angle 19°$

(C) $10 \text{ A}\angle 20°$

(D) $12 \text{ A}\angle 19°$

88. A balanced three-phase load has a power factor of 0.8 lagging. A DC resistance reading of one phase of the load shows 50 Ω of resistance. What is most nearly the reactance of the load?

(A) 31 Ω

(B) 35 Ω

(C) 38 Ω

(D) 52 Ω

89. The design for a transformer secondary calls for #14 gage copper wire with a 2.7 Ω/1000 ft resistance according to the NEC. The maximum current drawn by the electronic bias power drawer is 3 A, powered from the secondary of the transformer. The allowable voltage drop is 7.6 V. Most nearly, how far away from the transformer secondary can the power drawer be located?

(A) 140 m

(B) 290 m

(C) 470 m

(D) 1300 m

90. Which of the following statements regarding current transformers in protective systems is TRUE?

(A) Using a current transformer at higher frequencies makes saturation more likely.

(B) Using an instrument current transformer in place of a protective current transformer increases the time response.

(C) Maintenance should be performed only with the secondary of the current transformer shorted.

(D) both (A) and (C)

91. A permanent magnet instrument is represented graphically as follows.

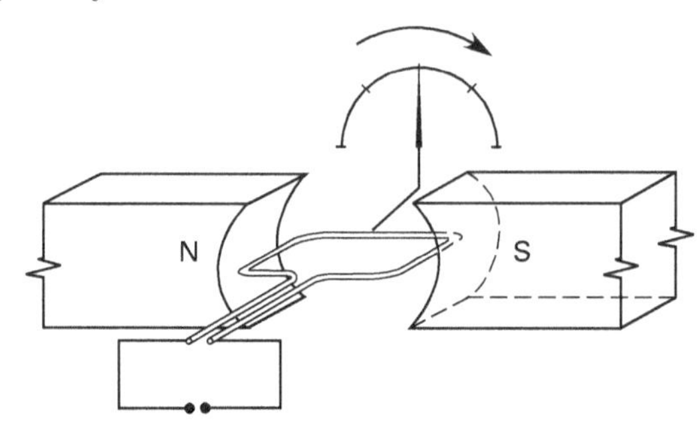

Which of the following sources causes clockwise rotation of the meter?

(A)

(B)

(C)

(D)

92. A synchronous generator is rated for 3 MVA, 11 kV, and has a power factor of 0.8. What is most nearly the magnitude of the rated current?

(A) 130 A

(B) 160 A

(C) 270 A

(D) 320 A

93. A Wheatstone bridge is connected as shown to measure an unknown resistance, R_x. The adjustable resistor is manipulated until V_{BD} equals 0 V. The indicator on the adjustable resistance shows 30 Ω at this point.

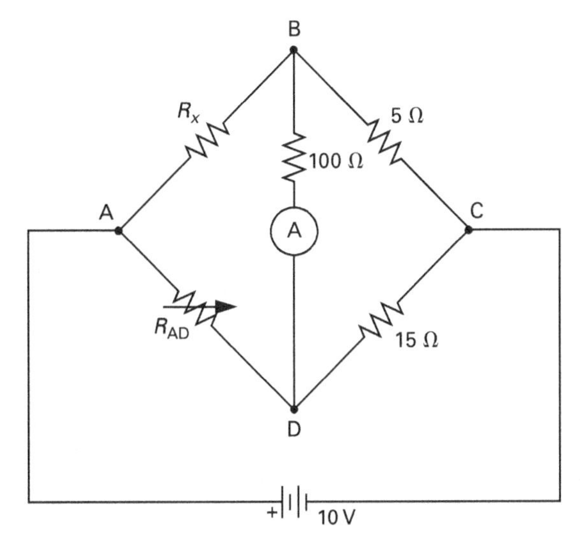

What is most nearly the value of the unknown resistance?

(A) 3 Ω

(B) 10 Ω

(C) 33 Ω

(D) 90 Ω

94. Consider the following one-line diagram of a three-phase distribution system with a fault at F1.

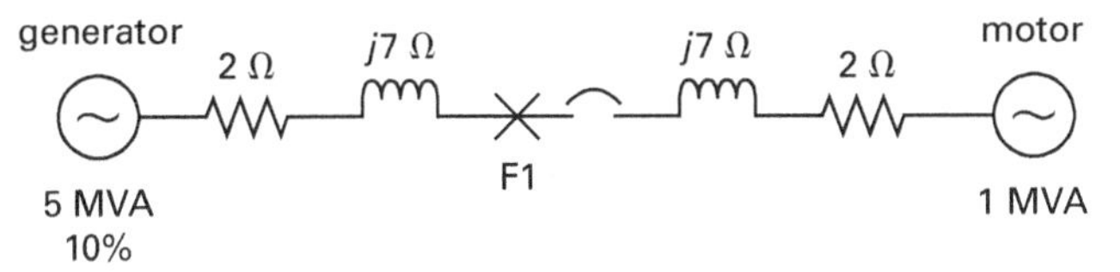

The motor is not online at the time of the fault. The transmission line is 13.8 kV. Using the MVA method, what is most nearly the short-circuit current at the fault?

(A) 0.55 kA

(B) 0.7 kA

(C) 1.3 kA

(D) 2.1 kA

95. Consider the signal shown. The signal is measured by two separate voltmeters. Voltmeter A is a DC voltmeter, and voltmeter B is a half-wave rectifier calibrated to read the rms value of a fully sinusoidal signal (that is, the output is scaled to that of a full-wave rectified signal).

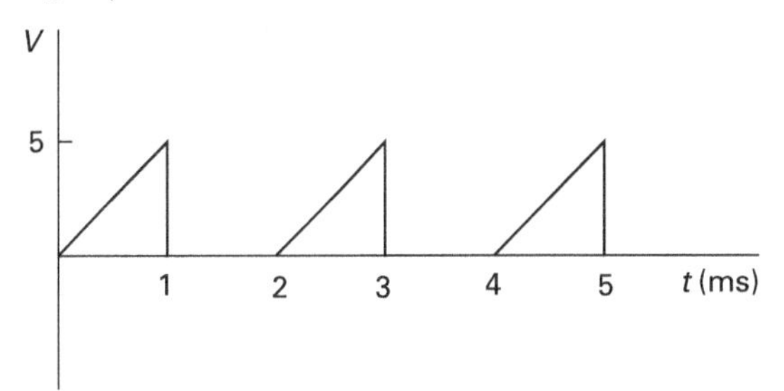

What is most nearly the ratio of the readings of voltmeter B to voltmeter A?

(A) 0.7

(B) 1.0

(C) 1.4

(D) 2.0

96. According to the NEC, a rod and pipe electrode must be buried up to ___ feet. If this is not possible, the electrode should be placed at a ___ degree angle or buried at least ___ inches.

(A) 2 ft, 30°, 12 in

(B) 2 ft, 45°, 24 in

(C) 4 ft, 45°, 24 in

(D) 8 ft, 45°, 30 in

97. Per NEC Sec. 110.6, conductors and cables are required to be marked either in their American Wire Gage (AWG) size or circular mil area. Conductor properties are shown in the following table. A plan written using the SI system calls for a conductor of 5.00 mm^2 or larger.

size (AWG)	area (cmil)
12	6530
10	10,380
8	16,510
4/0	211,600

For a conductor with circular cross-section, what is the minimum AWG size required?

(A) 12

(B) 10

(C) 8

(D) 4/0

98. A variable speed device intermittently fires Silicon Controlled Rectifiers (SCRs) to clip the current waveform, controlling the torque of the motor. Consider the following unity current waveform for a fully rectified signal.

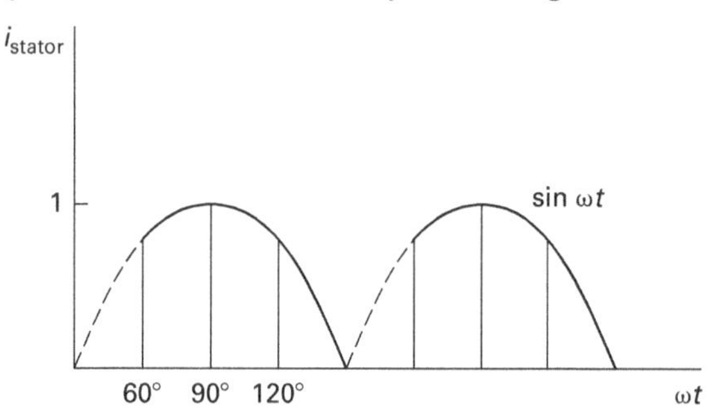

The average value of this waveform is $2/\pi$, or 0.637 (63.7%), of the waveform's peak value. It is desired to reduce the waveform to approximately 17% of its peak value. Most nearly, at what angle do the SCRs need to be programmed to fire to allow only 17% of the peak value to pass?

(A) 0°

(B) 60°

(C) 90°

(D) 120°

99. A medium voltage branch circuit is designed using AWG 10 THHN copper conductors. The expected ambient temperature is 50°C for the desert installation. The circuit also uses a switch whose terminals are rated for 75°C and a circuit breaker whose nameplate data indicates a 60°C terminal rating. NEC Table 310.15(B)(16) lists the applicable data.

What is most nearly the maximum permitted ampacity of the conductors given the indicated conditions?

(A) 17 A

(B) 30 A

(C) 32 A

(D) 40 A

100. According to the NEC, unless specifically permitted in another article, what is the maximum overcurrent protection for 12 AWG copper wire?

(A) 5.0 A

(B) 10 A

(C) 15 A

(D) 20 A

101. A branch circuit is to supply 20 A of continuous load and 5 A of noncontinuous load. Most nearly, what minimum conductor ampacity is required?

(A) 20 A

(B) 25 A

(C) 26 A

(D) 30 A

102. The following continuous single-phase loads are to be connected to the same circuit: a 5 hp motor rated at 230 V, and a 500 W, 240 V resistance heater. The ambient temperature is 30°C. According to the NEC, what is the minimum size of copper wire of type THHN, 90°C rated wiring required if installed in a raceway?

(A) 14 AWG

(B) 12 AWG

(C) 10 AWG

(D) 8 AWG

103. A branch circuit that does not contain receptacles is designed to carry 8 A of noncontinuous load and 20 A of continuous load. What standard size of overcurrent protection device (OCPD) is required?

(A) 15 A

(B) 30 A

(C) 35 A

(D) 40 A

104. A three-phase squirrel-cage Design B type induction motor is rated for 50 hp, 460 V, 60 Hz and has a 0.8 power factor. According to the NEC, what is most nearly the ratio of the locked-rotor current to the full-load current?

(A) 6.0

(B) 8.0

(C) 10

(D) 12

105. A circuit designed to carry 30 A at an ambient temperature of 30°C will use a four-conductor cable mounted in a raceway. The following AWG 10 copper cables are available for the installation: TW, THW, and THHW (90°C). NEC Table 310.15(B)(16) contains applicable data. What type of cable should be used to meet the required conditions, including any adjustment or derating factors?

(A) TW

(B) THW

(C) THHW (90°C)

(D) either THW or THHW (90°C)

106. From NEC Table 310.15(B)(16), the allowable ampacities for conductors are based on a variety of factors. Which of the following is NOT considered?

(A) all applicable ampacity factors

(B) compliance with requirements of product certifications

(C) preservation of safety benefits of established industry practices

(D) temperature compatibility within connected equipment

107. A certain transmission line has a measured capacitance of 100 pF/m and an inductance of 0.37 μH/m. What is most nearly the characteristic impedance?

(A) 0.00027 Ω

(B) 0.02 Ω

(C) 61 Ω

(D) 3700 Ω

108. The wye-connected ungrounded source has a phase-to-phase voltage of 12.5 kV. The source is connected to a balanced delta load grounded at the corner of phase C.

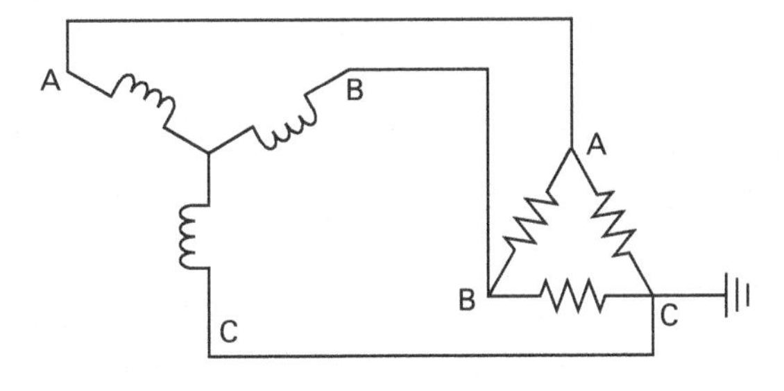

What is most nearly the voltage between the delta point A and the ground at point C?

(A) 0.0 kV

(B) 6.2 kV

(C) 7.2 kV

(D) 12 kV

109. A transmission line has a measured loss of 0.95 dB per 100 m. What is most nearly the loss in a 1 km line?

(A) 0.01 dB

(B) 0.95 dB

(C) 1.12 dB

(D) 9.50 dB

110. The balanced three-phase system shown has a balanced load. The source voltage and connection type, wye or delta, are unknown. The load voltage and line current are measured.

$$V_{AB} = 13.2 \text{ kV}\angle 0°$$
$$I_a = 50 \text{ A}\angle{-30°}$$

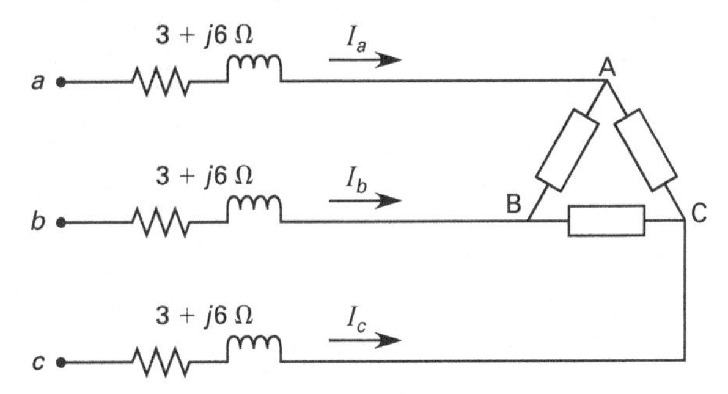

What is most nearly the magnitude of the source voltage, V_{ab}?

(A) 3.4 kV

(B) 7.9 kV

(C) 13 kV

(D) 23 kV

111. When using a Smith chart, the normalized load impedance must first be calculated to determine the correct location to place a compensating reactance. A transmission line has a characteristic impedance of 30 Ω and a load impedance of $120 + j150$ Ω. What is most nearly the normalized load impedance?

(A) $0.25 + j0.20$

(B) $3.00 + j4.00$

(C) $3.60 + j4.50$

(D) $4.00 + j5.00$

112. Direct current motors are used instead of alternating current motors in a variety of industrial applications. Which of the following is NOT a reason for using a DC motor instead of an AC motor?

(A) Speed control is simple and cost effective.

(B) Reversal is possible without power switching.

(C) DC motors stall, while AC motors can deliver five times the rated torque.

(D) Continuous operation over a speed range of 8:1 is possible.

113. A large commercial power plant uses a two-pole AC generator that produces a 60 Hz output. What is most nearly the rotational speed of the generator?

(A) 60 rpm

(B) 120 rpm

(C) 3600 rpm

(D) 7200 rpm

114. An inductor constructed as a solenoid with a cylindrical iron core is measured to have the following properties.

10 turns of wire

cross-sectional area, $A = 3.14\ \text{m}^2$

relative permeability, $\mu_r = 2000$

length, $l = 6$ in

The permeability of free space is 1.2566×10^{-6} H/m. What is most nearly the inductance of the solenoid with a 5 A current flowing?

(A) 5.2 H

(B) 8.2 H

(C) 13 H

(D) 26 H

115. A 1500 MW generator has a characteristic speed droop of 1.5%. The rated frequency is 60 Hz. The generator is online carrying 1000 MW at 60 Hz. What is most nearly the no-load frequency setpoint of the speed-governing system?

(A) 57.4 Hz

(B) 60.6 Hz

(C) 61.5 Hz

(D) 64.8 Hz

116. What produces ferromagnetism?

(A) electron spins in antiparallel pairs in closed shells

(B) exchange of atomic moments arranged in domains with equal alignment

(C) exchange of atomic moments with antiparallel arrangement of equal spins

(D) orbital or spin moments of electrons

117. Review the following power transmission one-line diagram.

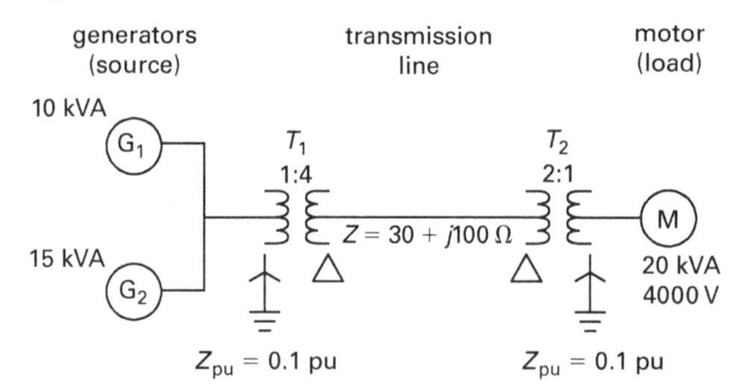

The line voltage at the output of generators 1 and 2 is selected as the base voltage. What is most nearly the base voltage?

(A) 1200 V

(B) 2000 V

(C) 2300 V

(D) 4000 V

118. The main distribution panel is supplied by a three-phase, 480 V system. The feeder supplying the propulsion power panel PP1 consists of three THWN, 300 kcmil copper conductors, and a neutral in an aluminum conduit measuring 500 m long. The balanced load draws 280 A at a lagging power factor of 0.8. According to the NEC, what is most nearly the voltage at panel PP1?

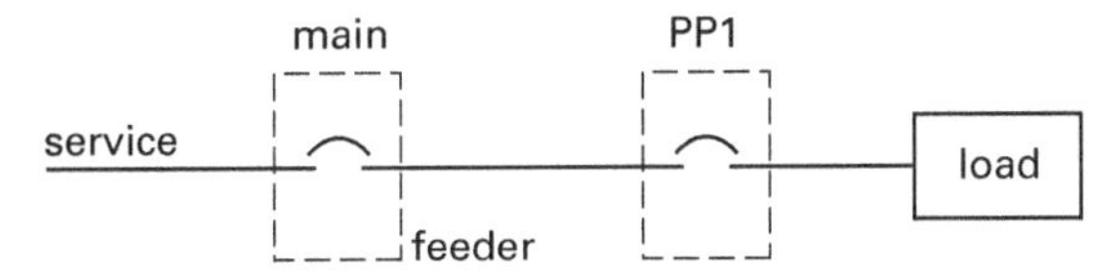

(A) 250 V

(B) 280 V

(C) 430 V

(D) 460 V

119. Consider the following three-phase load circuit.

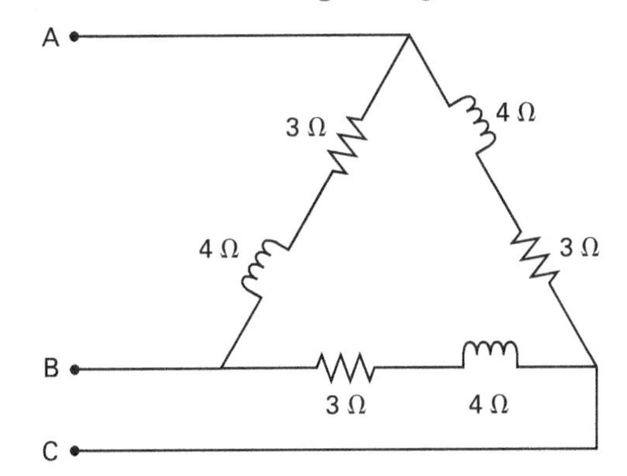

What is most nearly the phase impedance?

(A) $5\ \Omega\angle 53°$

(B) $5\ \Omega\angle 18°$

(C) $7\ \Omega\angle 25°$

(D) $15\ \Omega\angle 10°$

120. Which of the following represents a synchronous motor in an overexcited condition?

(A)

(B)

(C)

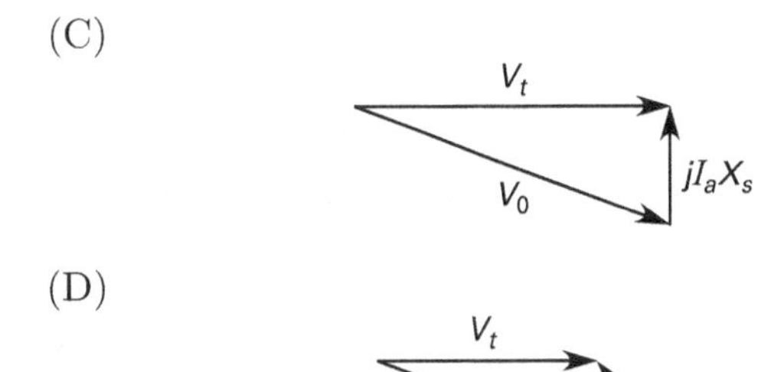

(D)

STOP!

DO NOT CONTINUE!

This concludes the Morning Session of the examination. If you finish early, check your work and make sure that you have followed all instructions. After checking your answers, you may submit your answers and leave the examination room. Once you leave, you will not be permitted to return to work or change your answers.

Afternoon Session Instructions

In accordance with the rules established by your state, you may use any approved battery- or solar-powered, silent calculator to work this examination. However, no blank papers, writing tablets, unbound scratch paper, or loose notes are permitted. Sufficient paper will be provided. The *NCEES PE Electrical Power Reference Handbook* and provided codes are the only references you are allowed to use during this exam.

You are not permitted to share or exchange materials with other examinees.

You will have four hours in which to work this session of the examination. Your score will be determined by the number of questions that you answer correctly. There is a total of 40 questions. All 40 questions must be worked correctly in order to receive full credit on the exam. There are no optional questions. Each question is worth 1 point. The maximum possible score for this section of the examination is 40 points.

Partial credit is not available. No credit will be given for methodology, assumptions, or work written on scratch paper.

Record all of your answers on the Answer Sheet. Mark your answers with a no. 2 pencil. Answers marked in pen may not be graded correctly. Marks must be dark and must completely fill the bubbles. Record only one answer per question. If you mark more than one answer, you will not receive credit for the question. If you change an answer, be sure the old bubble is erased completely; incomplete erasures may be misinterpreted as answers.

If you finish early, check your work and make sure that you have followed all instructions. After checking your answers, you may submit your answers and leave the examination room. Once you leave, you will not be permitted to return to work or change your answers.

When permission has been given by your proctor, you may begin your examination.

Do not work any questions from the Morning Session during the second four hours of this exam.

Name: ____
Last First Middle Initial

Examinee number: ______________

Examination Booklet number: ______________

Principles and Practice of Engineering Examination

Afternoon Session Practice Exam 2

Afternoon Session

121. (A) (B) (C) (D)	131. (A) (B) (C) (D)	141. (A) (B) (C) (D)	151. (A) (B) (C) (D)
122. (A) (B) (C) (D)	132. (A) (B) (C) (D)	142. (A) (B) (C) (D)	152. (A) (B) (C) (D)
123. (A) (B) (C) (D)	133. (A) (B) (C) (D)	143. (A) (B) (C) (D)	153. (A) (B) (C) (D)
124. (A) (B) (C) (D)	134. (A) (B) (C) (D)	144. (A) (B) (C) (D)	154. (A) (B) (C) (D)
125. (A) (B) (C) (D)	135. (A) (B) (C) (D)	145. (A) (B) (C) (D)	155. (A) (B) (C) (D)
126. (A) (B) (C) (D)	136. (A) (B) (C) (D)	146. (A) (B) (C) (D)	156. (A) (B) (C) (D)
127. (A) (B) (C) (D)	137. (A) (B) (C) (D)	147. (A) (B) (C) (D)	157. (A) (B) (C) (D)
128. (A) (B) (C) (D)	138. (A) (B) (C) (D)	148. (A) (B) (C) (D)	158. (A) (B) (C) (D)
129. (A) (B) (C) (D)	139. (A) (B) (C) (D)	149. (A) (B) (C) (D)	159. (A) (B) (C) (D)
130. (A) (B) (C) (D)	140. (A) (B) (C) (D)	150. (A) (B) (C) (D)	160. (A) (B) (C) (D)

Exam 2: Afternoon Session

121. Of the four equivalent circuits shown, which represents an induction motor?

(A)

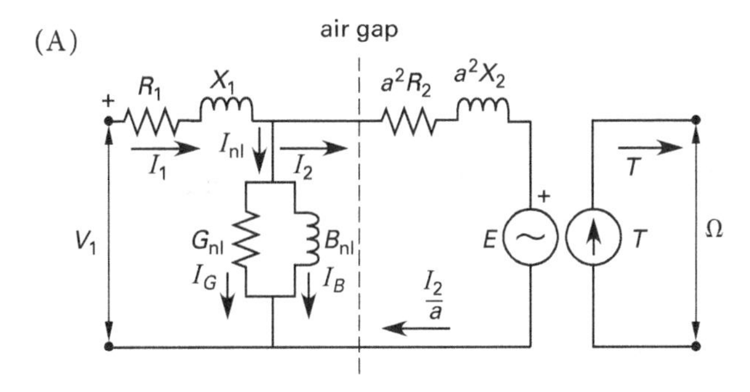

(B)

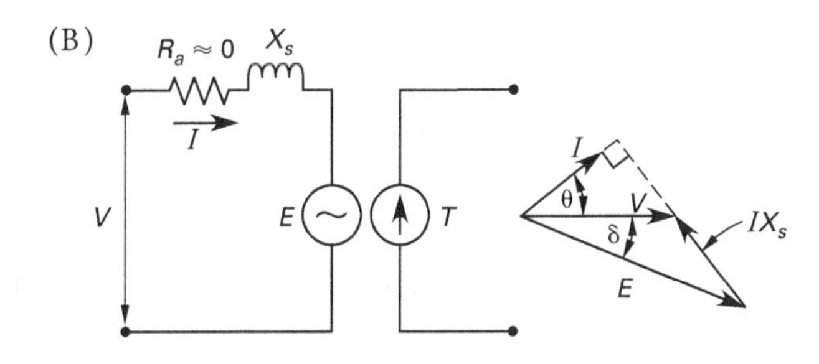

(C)

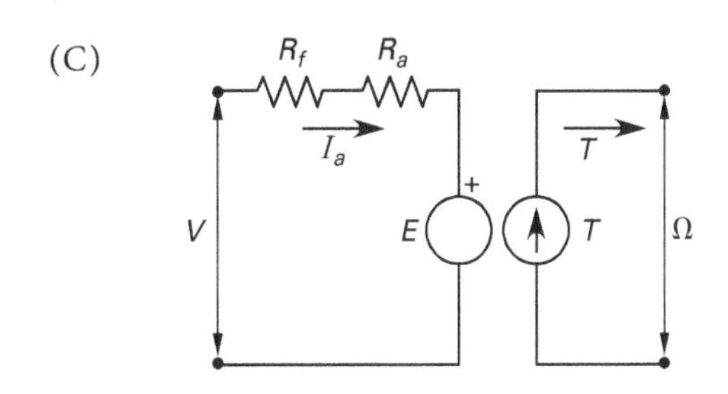

(D)

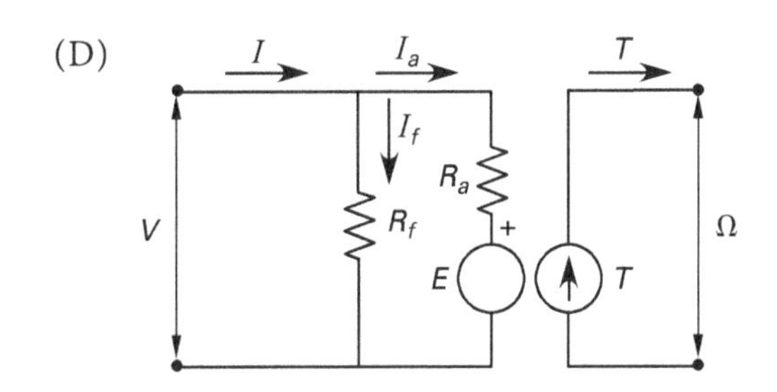

122. Consider the Wheatstone bridge shown.

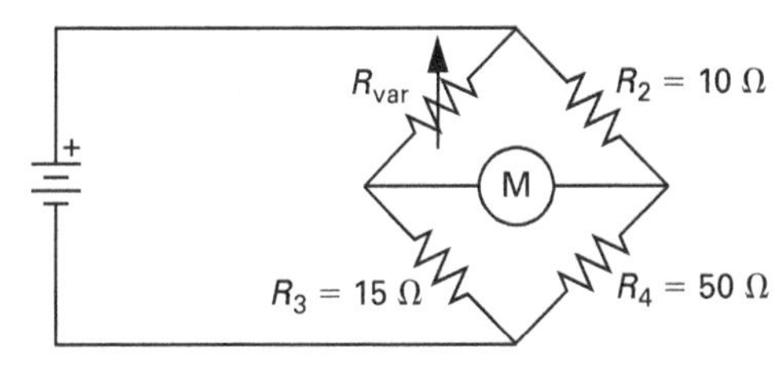

What value of the variable resistor, R_{var}, is most nearly required to give a zero indication (i.e., input current of zero) on the meter?

(A) 3.0 Ω

(B) 33 Ω

(C) 36 Ω

(D) 75 Ω

123. An induction motor producing 10 hp is connected to a three-phase, 240 V, 60 Hz power line. The stator windings are connected in a wye configuration. The synchronous speed is 1800 rpm, and the motor full-load speed is 1740 rpm. The motor power factor is 0.70, and the energy efficiency is 80%. What is most nearly the slip percentage?

(A) 3.3%

(B) 3.4%

(C) 20%

(D) 30%

124. A digital multimeter has four voltage scales, and its environmental specifications require operation at 25°C. The meter responds to signals from 0 Hz to 1000 Hz. The Boltzmann constant is 1.3807×10^{-23} J/K. What is most nearly the thermal agitation noise on the multimeter's 20 kΩ resistor?

(A) 1.7×10^{-10} V

(B) 1.8×10^{-8} V

(C) 5.2×10^{-7} V

(D) 5.7×10^{-7} V

125. A 15 hp motor is tested, and its full-load losses are presented in the table. The motor operates at an rms voltage of 240 V and a 60 Hz frequency.

type of loss	magnitude (W)
stator core	650
stator copper	500
rotor copper	150
friction and windage	950

What is most nearly the efficiency of the motor?

(A) 20%

(B) 40%

(C) 80%

(D) 90%

126. Which of the following statements about variable frequency drive (VFD) controllers for an induction motor is FALSE?

(A) Insulated gate bipolar junction transistors are often used in VFD controllers.

(B) VFD controllers start a motor using low frequency.

(C) Harmonics is a negative side effect of VFDs.

(D) The volts/hertz ratio is variable in order to deliver constant torque.

127. Consider the magnetic hysteresis curve shown.

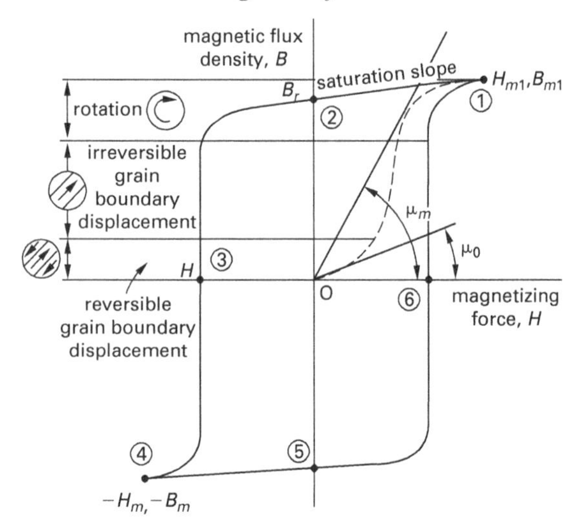

The coercive force is from the origin, O, to what point?

(A) point 1

(B) point 2

(C) point 3

(D) point 4

128. A cast-iron toroid is used in a magnetic circuit. The circuit requires a flux of 0.01 Wb for proper operation. The cast-iron toroid has a relative permeability of 2000. The constant cross-sectional area of the flux path is 0.0045 m^2 with a diameter of 0.076 m.

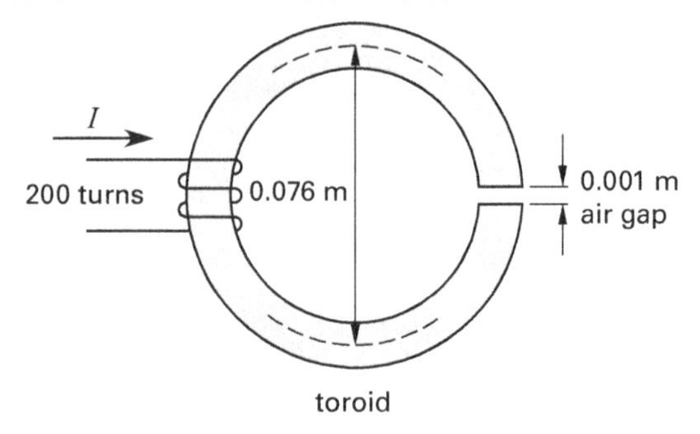

What is most nearly the current required for proper operation?

(A) 1 A

(B) 9 A

(C) 10 A

(D) 20 A

129. A protective device called a ground-fault circuit interrupter (GFCI) is required in certain locations to protect personnel. Which of the following statements relating to GFCIs is FALSE?

(A) The approximate magnetic flux within a GFCI is zero during normal operation.

(B) The GFCI is designed to protect against inadvertently grounded equipment.

(C) A GFCI operates on current ranging from 4 mA to 6 mA.

(D) A GFCI is designed to protect against a phase-to-neutral short circuit.

130. A nameplate on a 115 V shaded-pole motor specifies $^1/_4$ hp and a 6.0 A full-load current. According to NEC Art. 430, what current should be used to determine the correct branch-circuit conductors?

(A) 3.1 A

(B) 5.8 A

(C) 6.0 A

(D) 6.4 A

131. As a result of their high frequency, high-frequency transmission lines undergo an effect that causes AC resistance to increase and internal inductance to decrease when compared to lower frequency circuits. What is the name of this effect?

(A) infrared effect

(B) radiated resistance

(C) skin effect

(D) space propagation loss

132. An AC time overcurrent relay is used to monitor an 11 kV transmission line. (The relay is designated 51 by IEEE Standard C37.2, *IEEE Standard Electrical Power System Device Function Numbers and Contact Designations.*) The minimum pickup of the relay is 333 A. A three-phase short on this line draws 2000 A. Based on the relay characteristics shown and a desired coordination time of 0.5 s, what is the time dial setting required for the desired operation?

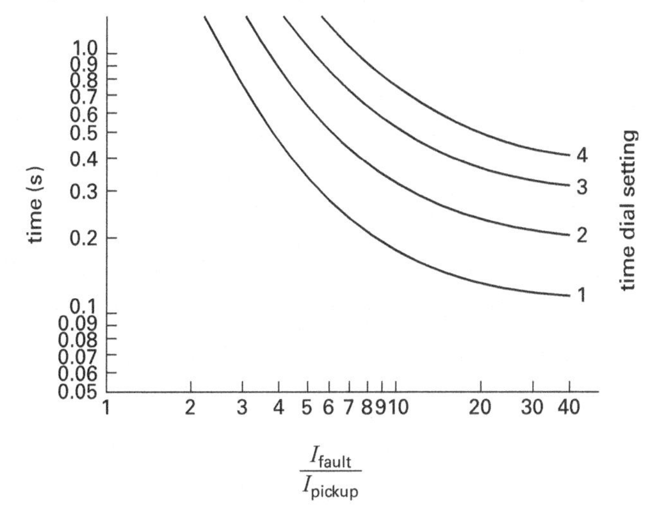

(A) 1

(B) 2

(C) 3

(D) 4

133. What is most nearly the characteristic impedance of a high-frequency transmission line with $Z_l = 0.85\ \Omega/\text{m}$ and $Y_l = 7.00 \times 10^{-6}$ S/m?

(A) 2.5 mΩ

(B) 1.0 Ω

(C) 350 Ω

(D) 12 MΩ

134. A fault has occurred on the following single-phase circuit. There are no sources downstream of the fault. When the fault resistance is 0 Ω, the fault current is 10,000 A.

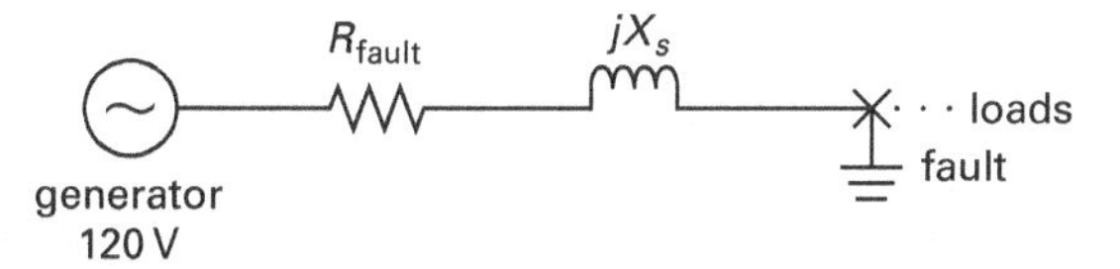

What is most nearly the fault current when the fault resistance is 0.5 Ω?

(A) 240 A

(B) 550 A

(C) 2500 A

(D) 15,000 A

135. The standing-wave ratio (SWR) of a given transmission line is measured as 2. What is most nearly the fraction of incident power reflected back to the source?

(A) 0.1

(B) 0.3

(C) 0.5

(D) 2

136. An ideal single-phase transformer has a primary resistance of 5 Ω. The secondary resistance is 7 Ω, and the secondary reactance is 7 Ω.

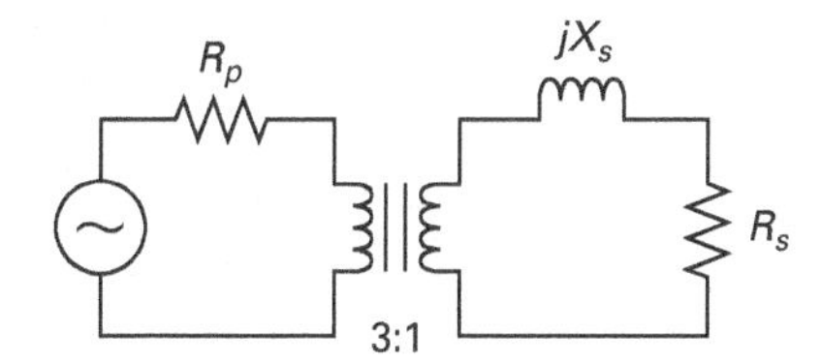

What is most nearly the effective primary impedance?

(A) 9.0 Ω∠55°

(B) 89 Ω∠45°

(C) 90 Ω∠43°

(D) 90 Ω∠55°

137. For the three-phase, 13.2 kV transmission line shown, the source impedance is 10 $\Omega\angle 85°$. The impedance between station A and station B, Z_{AB}, is 15 $\Omega\angle 75°$. The impedance between station B and station C, Z_{BC}, is 12 $\Omega\angle 70°$. Assume a fault occurs at station C.

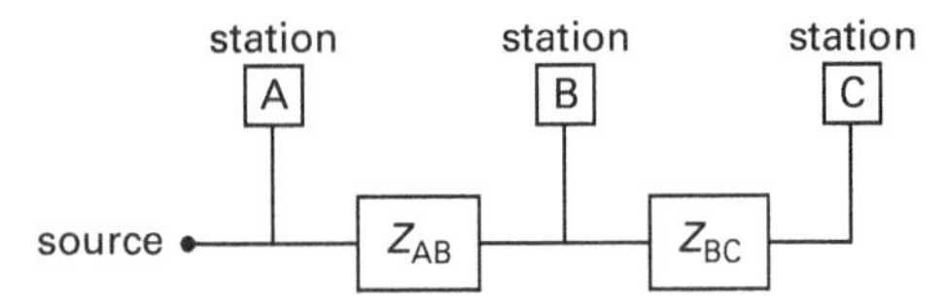

What is most nearly the magnitude of a three-phase fault current at station C?

(A) 210 A

(B) 310 A

(C) 350 A

(D) 550 A

138. A 20 hp, 208 V, three-phase induction motor has a service factor of 1.0 and is rated for continuous duty. A separate fused overload device is to be installed. According to the NEC, which standard rated fuse is the best choice?

(A) 50 A

(B) 60 A

(C) 65 A

(D) 70 A

139. A protection system diagram using IEEE Standard C37.2 shows a contact with the following coding.

$$\frac{27-2}{a}$$

The device is best described as an

(A) undervoltage relay normally closed contact

(B) undervoltage relay normally open contact

(C) overcurrent relay secondary contact

(D) overcurrent relay auxiliary contact

140. A portion of a two-generator three-phase distributor is shown as a one-line diagram.

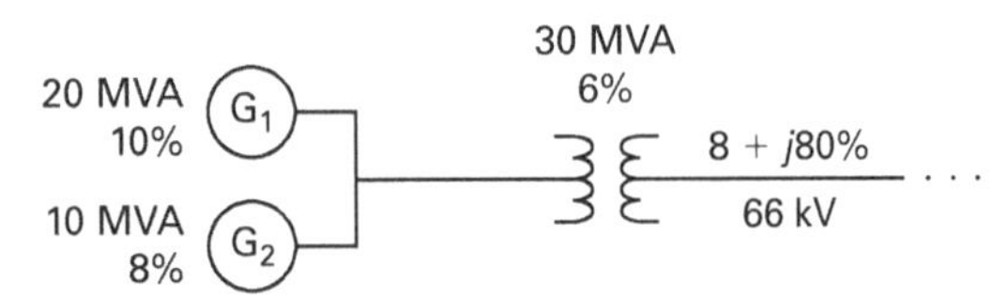

Which of the following impedance diagrams represents the distribution system using a 20 MVA base?

(A)

G_1 10%, G_2 8%, 18%, 8 + j80%

(B)

G_1 1%, G_2 0.8%, 0.9%, 0.8 + j8%

(C)

G_1 10%, G_2 10%, 9%, 3.7 + j37%

(D)

G_1 10%, G_2 16%, 4%, 3.7 + j37%

141. A one-line diagram of a three-phase distribution system is shown.

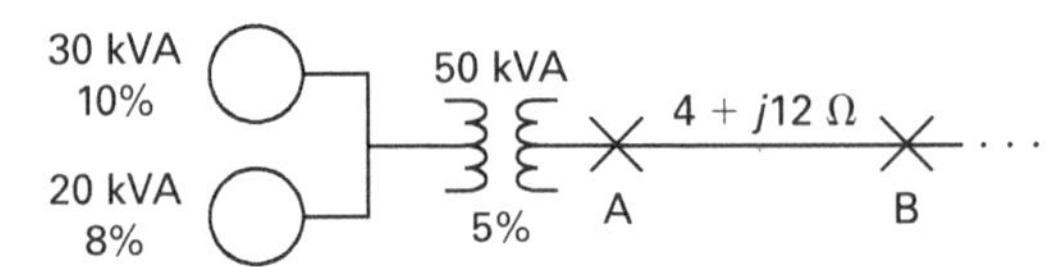

What is most nearly the short-circuit power if a fault occurs at point A?

(A) 50 kVA

(B) 90 kVA

(C) 400 kVA

(D) 900 kVA

142. The following one-line diagram shows a three-phase system. A fault occurs directly at the outlet of the transformer.

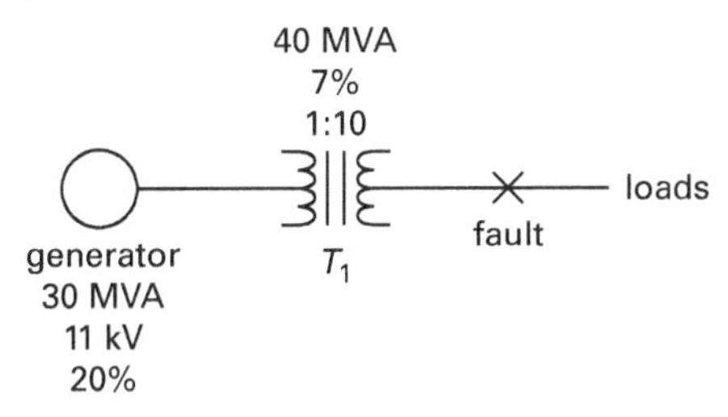

What is most nearly the apparent power contribution to the fault from the generator?

(A) 30 MVA

(B) 120 MVA

(C) 150 MVA

(D) 620 MVA

143. From the Wien displacement law (with a constant of 2.8978×10^{-3} m·K), what wavelength will radiate the most power from a light source at 2900K?

(A) 9.9×10^{-7} m

(B) 3.3×10^{-4} m

(C) 2.9×10^{-3} m

(D) 3.0×10^{-3} m

144. Power flow studies use different buses for reference. Which of the following buses is used as the reference for the voltage angle?

(A) generator bus

(B) load bus

(C) slack bus

(D) voltage-controlled bus

145. An office space will use 120 V general-purpose receptacle outlets of the following type and number: 45 duplex, 5 triplex, and 10 quad. This feeder/service load is considered separately from the lighting load when applying demand factors. What is most nearly the total calculated load?

(A) 94 A

(B) 96 A

(C) 105 A

(D) 109 A

146. What is the zero-sequence model for a delta-delta transformer?

(A)

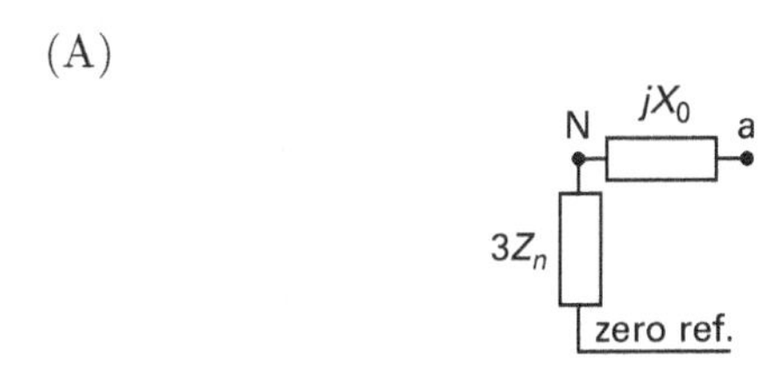

(B)

A B
Z
zero ref.

(C)

A Z B
zero ref.

(D)

A Z B
zero ref.

147. A three-phase, alternating-current squirrel-cage motor's nameplate data provides the following information.

power	25 hp
service factor	1.00
temperature rise	60°C
full-load rating	68 A
voltage	230 V
locked rotor	D

Which overload setting should be selected?

(A) 60 A

(B) 70 A

(C) 80 A

(D) 90 A

148. An underfrequency condition in an electrical system

(A) may result in thermal damage to cores

(B) causes motors to draw heavy overloads

(C) results from a loss of excitation condition on the generator

(D) results in reduced ventilation and affects the turbine more than the generator

149. LEDs are to be installed in a hazardous area. They will be installed as simple apparatus in a standard 40°C ambient environment. The parameters of the LEDs from the installation specifications are shown.

$$P_{\text{out}} = 19.5\ \text{W}$$

$$R_{\text{thermal}} = 9.6\ \frac{°\text{C}}{\text{W}}$$

Which temperature code (or class) is associated with this LED installation?

(A) T2C

(B) T2D

(C) T3

(D) T6

150. Which term describes a device that introduces inductive or capacitive reactance into an electrical circuit?

(A) capacitor

(B) coil

(C) reactor

(D) all of the above

151. Review the following power transmission one-line diagram.

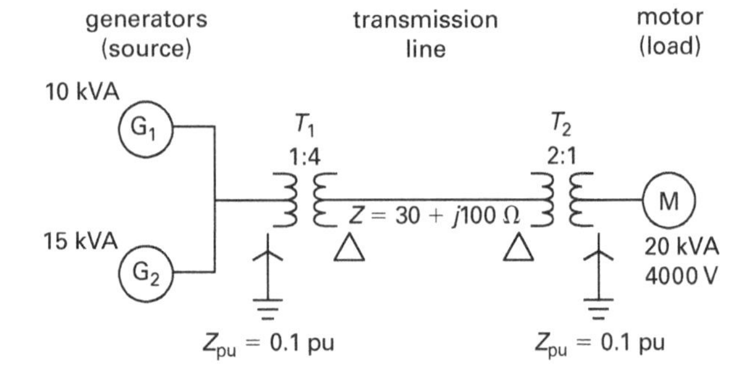

What is most nearly the per-unit impedance of the transmission line assuming a base voltage of 2 kV and a base power of 10 kVA?

(A) $0.012 + j0.025$ pu

(B) $0.030 + j0.100$ pu

(C) $0.075 + j0.250$ pu

(D) $0.150 + j0.500$ pu

152. A 20 MVA generator supplies loads via a 2 MVA transformer. The transformer has a per-unit impedance of 5% and a secondary voltage of 460 V. The three-phase system X/R ratio is the same for the generator and the transformer. What is most nearly the short-circuit current available for a fault downstream of the transformer?

(A) 10 kA

(B) 20 kA

(C) 30 kA

(D) 50 kA

153. Consider the following three-phase load circuit.

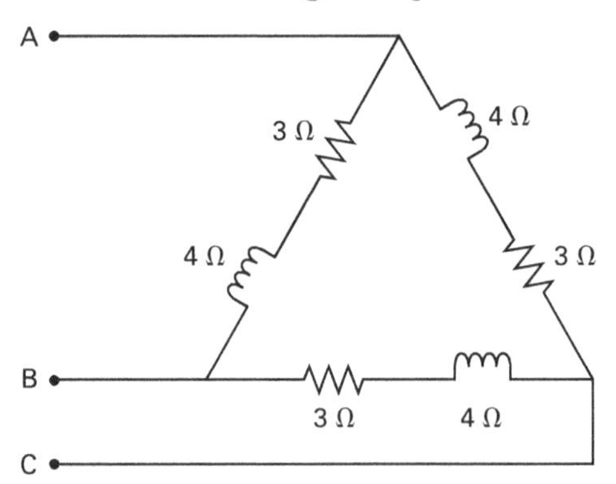

If the line voltage is 120 V, what is most nearly the phase voltage?

(A) 7.0 V

(B) 25 V

(C) 70 V

(D) 120 V

154. Which of the following statements regarding AC and DC rotating machines with equal output power is FALSE?

(A) Low-speed machines are larger than high-speed machines.

(B) Physical size depends upon the speed of rotation.

(C) Physical size depends upon the power.

(D) The size of the machine is limited by temperature.

155. Consider the following single-phase transformer.

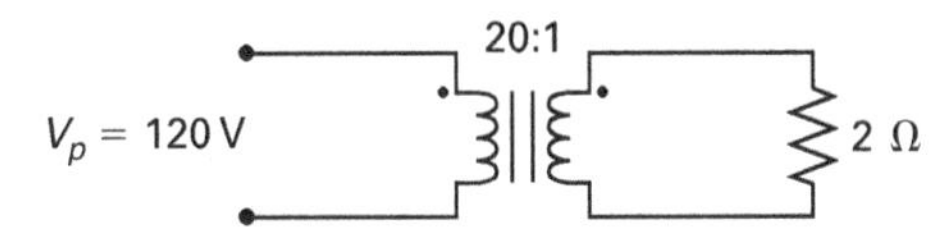

What is most nearly the current through the load resistor?

(A) 3 A

(B) 6 A

(C) 10 A

(D) 60 A

156. A 12 V DC battery has an internal resistance of 0.1 Ω. A starter motor draws 4 A. Ignoring wiring resistance, what is most nearly the voltage at the starter terminal?

(A) 0.4 V

(B) 1.2 V

(C) 4.0 V

(D) 12 V

157. A six-pole, 10 hp induction motor rated at 90% efficiency operates on a 60 Hz line. What input power is most nearly required at full load?

(A) 10 W

(B) 6700 W

(C) 7500 W

(D) 8300 W

158. A 200 hp, 250 VA synchronous motor is operating at 0.8 pf leading in parallel with a 50 hp, 75 VA induction motor operating at 0.7 pf lagging. What is most nearly the system power factor?

(A) 0.25 leading

(B) 0.75 leading

(C) 0.75 lagging

(D) 0.90 leading

159. What type of DC motor produces the greatest starting torque?

(A) compound

(B) series

(C) shunt

(D) wound rotor

160. Which of the following signals belongs to the indicated circuit?

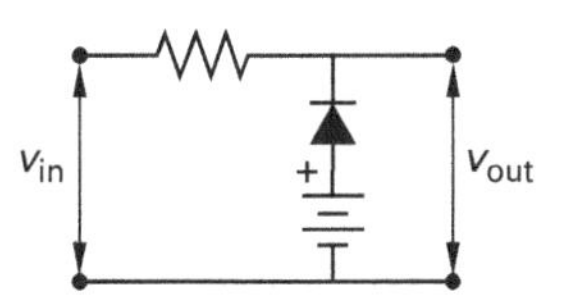

(A)

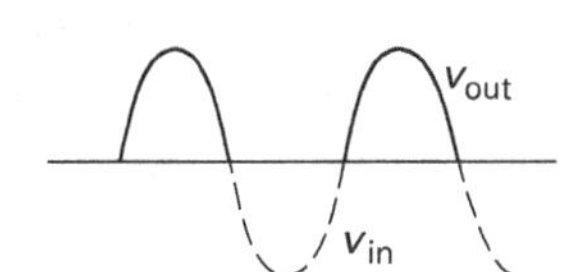

(B)

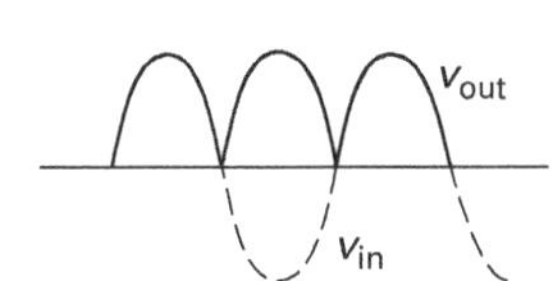

(C)

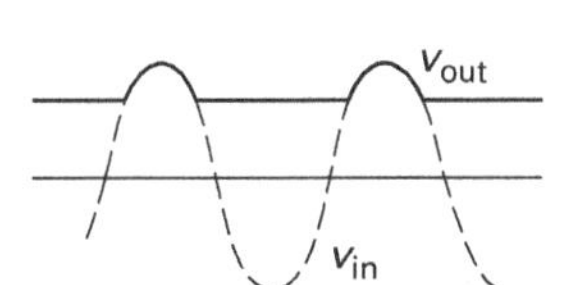

(D)

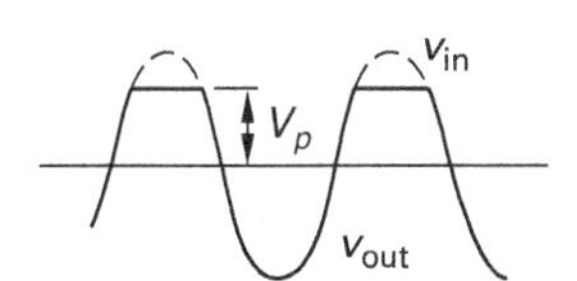

STOP!

DO NOT CONTINUE!

This concludes the Afternoon Session of the examination. If you finish early, check your work and make sure that you have followed all instructions. After checking your answers, you may submit your answers and leave the examination room. Once you leave, you will not be permitted to return to work or change your answers.

Exam 1 Answer Keys

Morning Session Answer Key

No.	Answer	No.	Answer	No.	Answer	No.	Answer
1.	D	11.	C	21.	C	31.	C
2.	C	12.	A	22.	A	32.	D
3.	A	13.	D	23.	B	33.	C
4.	A	14.	B	24.	D	34.	B
5.	B	15.	B	25.	B	35.	D
6.	D	16.	A	26.	D	36.	B
7.	B	17.	D	27.	A	37.	B
8.	B	18.	D	28.	B	38.	D
9.	B	19.	C	29.	B	39.	D
10.	B	20.	C	30.	A	40.	B

Afternoon Session Answer Key

No.	Answer	No.	Answer	No.	Answer	No.	Answer
41.	D	51.	D	61.	D	71.	A
42.	B	52.	D	62.	A	72.	C
43.	A	53.	B	63.	D	73.	B
44.	C	54.	A	64.	A	74.	D
45.	C	55.	A	65.	A	75.	B
46.	C	56.	B	66.	A	76.	A
47.	A	57.	C	67.	C	77.	C
48.	A	58.	D	68.	C	78.	C
49.	D	59.	D	69.	D	79.	C
50.	A	60.	A	70.	C	80.	D

Exam 2 Answer Keys

Morning Session Answer Key

No.	Answer	No.	Answer	No.	Answer	No.	Answer
81.	A	91.	B	101.	D	111.	D
82.	B	92.	B	102.	C	112.	C
83.	C	93.	B	103.	C	113.	C
84.	A	94.	B	104.	A	114.	A
85.	B	95.	D	105.	C	115.	B
86.	B	96.	D	106.	A	116.	B
87.	B	97.	B	107.	C	117.	B
88.	C	98.	D	108.	D	118.	C
89.	A	99.	B	109.	D	119.	A
90.	C	100.	D	110.	C	120.	D

Afternoon Session Answer Key

No.	Answer	No.	Answer	No.	Answer	No.	Answer
121.	A	131.	C	141.	C	151.	C
122.	A	132.	B	142.	B	152.	B
123.	A	133.	C	143.	A	153.	D
124.	D	134.	A	144.	C	154.	D
125.	C	135.	A	145.	A	155.	A
126.	D	136.	C	146.	D	156.	D
127.	C	137.	A	147.	B	157.	D
128.	C	138.	B	148.	D	158.	D
129.	D	139.	B	149.	A	159.	B
130.	C	140.	D	150.	C	160.	C

Solutions

Exam 1: Morning Session

Content in blue refers to the *NCEES Handbook*.
Content in red is additional essential information.

1. The per-unit system can be referenced to phase or line values. The line values are given here since there is no indication of the internal wiring, wye or delta, of the generators or the motor. The result in either the three-phase base (i.e., the line values base) or the per-phase base (i.e., the phase value base) would be identical.

The power used is the apparent power, as the lines and transformers must be designed to handle this power. The per-unit power of the motor is

$$S_{\text{kVA,pu}} = \frac{S_{\text{kVA,motor}}}{S_{\text{kVA,base}}} = \frac{20\ \text{kVA}}{10\ \text{kVA}}$$
$$= 2.0\ \text{pu}$$

The answer is (D).

2. High-resistance neutral or ground connections on the wye perform protective functions unrelated to lightning strikes. This eliminates option A and option B.

Placing surge protectors on the transformer secondary side will not protect the primary side. This eliminates option D.

A phase-to-ground connection is required to shunt the energy of a lightning strike. On the primary side, the delta connection has no such ground connection and thus would require one, with a surge arrester between the phase and ground, for protection.

The answer is (C).

3. The applicable formula is

$$P = IV$$

Rearranging to solve for V,

$$V = \frac{P}{I} = \frac{\left(10.8\ \frac{\text{kJ}}{\text{min}}\right)\left(1000\ \frac{\text{J}}{\text{kJ}}\right)}{900\ \frac{\text{C}}{\text{min}}}$$
$$= 12\ \text{V}$$

The answer is (A).

4. The actual impedance is

$$Z_{\text{pu}} = \frac{Z_{\text{actual}}}{Z_{\text{base}}}$$
$$Z_{\text{actual}} = Z_{\text{pu}} Z_{\text{base}}$$

The base impedance is unknown in this equation. To obtain the base impedance for a three-phase system, calculate

Per Unit System

$$Z_{\text{base}} = \frac{V_{\text{base}}}{\sqrt{3}\, I_{\text{base}}}$$

The base current is unknown in this equation. The turns ratio, a, is given as 25. The base voltage, which is rated voltage on the secondary side, is

Single-Phase Transformer Equivalent Circuits

$$a = \frac{V_p}{V_s}$$
$$V_s = \frac{V_p}{a} = \frac{11{,}000\ \text{V}}{25}$$
$$= 440\ \text{V}$$

In the per-unit system, naming any two of the bases sets the other two. The base power is the rated power of 3600 kVA, while the base voltage on the secondary side is the rated voltage of 440 V. The base current, the unknown value in the base impedance equation, is

Per Unit System

$$I_{\text{base}} = \frac{S}{\sqrt{3}\, V}$$
$$= \frac{(3600\ \text{kVA})\left(1000\ \frac{\text{VA}}{\text{kVA}}\right)}{\sqrt{3}\,(440\ \text{V})}$$
$$= 4724\ \text{A}$$

Substitute this value into the equation for the base impedance.

$$Z_{base} = \frac{V_{base}}{\sqrt{3}\,I_{base}} = \frac{440\ \text{V}}{\sqrt{3}(4724\ \text{A})} = 0.0537\ \Omega$$

The base impedance could have been calculated more directly with the following.

Per Unit System

$$Z_{base} = \frac{V_{base}^2}{S_{base}} = \frac{(440\ \text{V})^2}{3600 \times 10^3\ \text{VA}} = 0.0537\ \Omega$$

Substitute this value into the equation for the actual impedance.

$$Z_{actual} = Z_{pu}Z_{base} = (0.075)(0.0537\ \Omega) = 0.004\ \Omega$$

The answer is (A).

5. The impedance of each circuit component is determined as follows.

Phasor Transforms of Sinusoids

$$Z_R = 50\ \Omega\angle 0°$$

$$Z_C = \frac{1}{j\omega C} = \frac{1}{\omega C}\angle -90° = \frac{1}{(377\ \text{Hz})(50 \times 10^{-6}\ \text{F})}\angle -90° = 53.05\ \Omega\angle -90°$$

$$Z_L = j\omega L = \omega L\angle +90° = (377\ \text{Hz})(50 \times 10^{-3}\ \text{H})\angle 90° = 18.85\ \Omega\angle 90°$$

Combine the resistor and capacitor impedances in a parallel manner.

$$Z_R \| Z_C = \frac{Z_R Z_C}{Z_R + Z_C} = \frac{(50\ \Omega\angle 0°)(53.05\ \Omega\angle -90°)}{50 - j53.05\ \Omega} = 26.48 - j24.95\ \Omega = 36.39\ \Omega\angle -43.3°$$

The resistor-capacitor combination is also parallel with the inductor. Use the following formula to determine the equivalent impedance of the circuit.

$$Z_{equiv} = (Z_R \| Z_C) \| Z_L = \frac{(Z_R \| Z_C)Z_L}{Z_R \| Z_C + Z_L} = \frac{(36.39\ \Omega\angle -43.3°)(18.85\ \Omega\angle 90°)}{(26.48 - j24.95\ \Omega) + j18.85\ \Omega} = 25.23\ \Omega\angle 59.67° \quad (25\ \Omega\angle 60°)$$

The answer is (B).

6. Option A defines the transient insulation level. Option B defines the lightning impulse insulation level. Option C defines the basic switching impulse insulation level. Option D defines the basic lightning impulse insulation level.

The answer is (D).

7. Treat the voltage sources as short circuits.

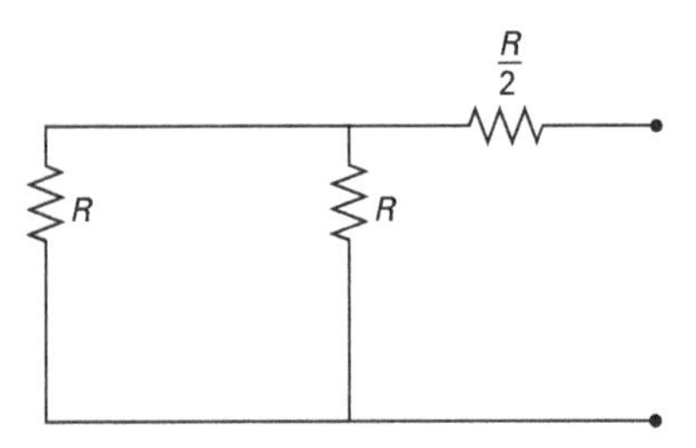

Combine the resistors in parallel.

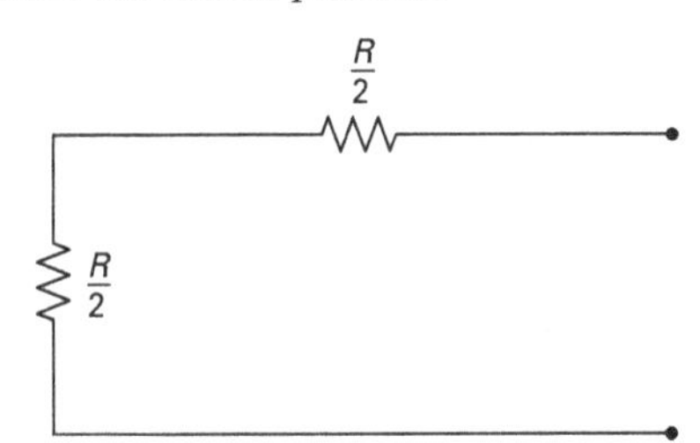

Combine the resulting series resistors to obtain the Thevenin equivalent resistance.

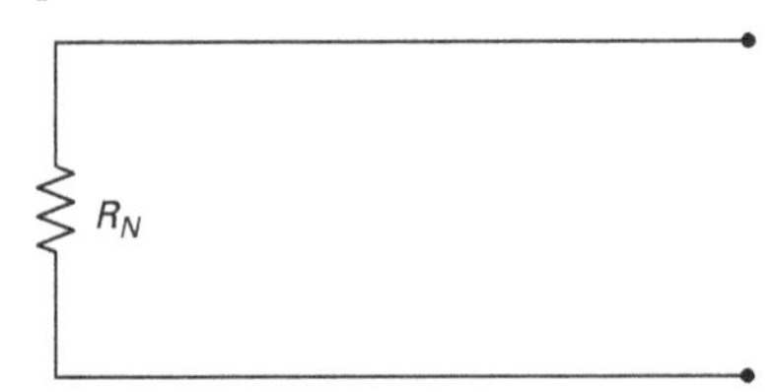

The answer is (B).

8. For this unbalanced load, the following condition must exist.

$$\mathbf{I}_A + \mathbf{I}_C + \mathbf{I}_N = 0$$

The phase voltage is

3-Phase Circuits

$$V_p = \frac{V_l}{\sqrt{3}} = \frac{480\text{ V}}{\sqrt{3}} = 277\text{ V}$$

Rearrange to solve for the neutral current, and then take the magnitude of that current. The neutral current is

$$\begin{aligned}\mathbf{I}_A + \mathbf{I}_C + \mathbf{I}_N &= 0 \\ |-\mathbf{I}_N| &= |\mathbf{I}_A + \mathbf{I}_C| \\ I_N &= \left|\frac{S_A}{\mathbf{V}_{\phi A}} + \frac{S_C}{\mathbf{V}_{\phi C}}\right| \\ &= \left|\frac{500 + j300\text{ VA}}{277\text{ V}\angle 0°} + \frac{500 + j300\text{ VA}}{277\text{ V}\angle -240°}\right| \\ &= |2.1\text{ A}\angle 31° + 2.1\text{ A}\angle -89°| \\ &= |2.1\text{ A}\angle 29°| \\ &= 2.1\text{ A}\end{aligned}$$

The answer is (B).

9. At time $t = 0$ s, the circuit is as shown.

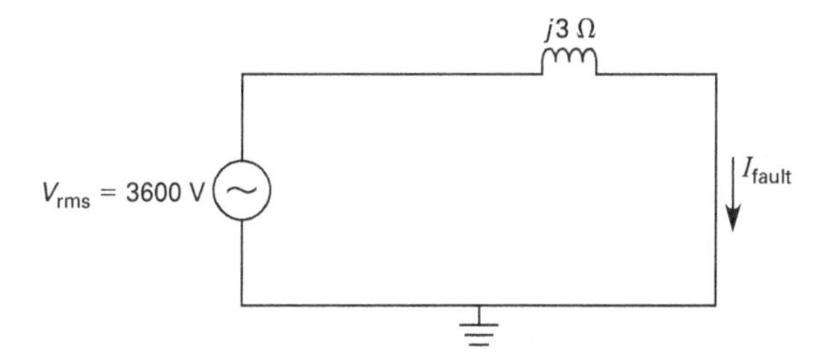

The rms fault current is

$$I_{\text{fault}} = \frac{V}{Z} = \frac{3600\text{ V}}{3\ \Omega} = 1200\text{ A}$$

The answer is (B).

10.

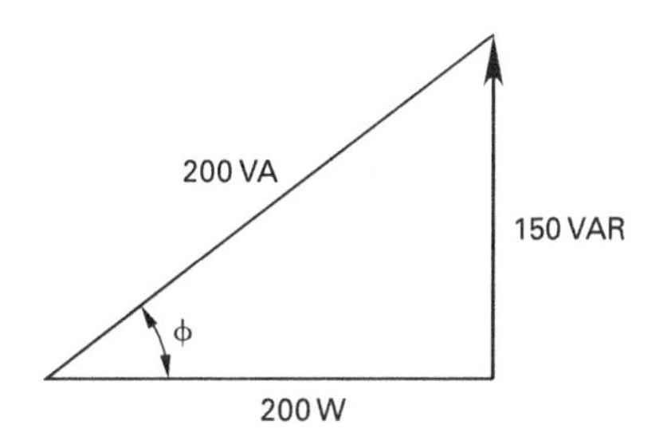

The phrase "correct the power factor" indicates a desire to make the load totally resistive, that is, to change the power factor to a value of 1.0. The given load consists of 200 W of real load and 150 VAR of reactive load (inductive in this case, given the positive value). Therefore, the capacitance must correct the power factor by providing an opposing −150 VAR. Since this is a three-phase system, each phase will correct for one-third of the total reactive power, or −50 VAR. The negative sign indicates the capacitive nature of the load. First, calculate the capacitive reactance required to generate the −50 VAR.

The reactive power is as follows. The final value is for a purely reactive load. In some calculations, the magnitude of the current is replaced by $\mathbf{I}_r$, meaning the reactive portion of the current.

Phasor Transforms of Sinusoids

$$\mathbf{Q} = \mathbf{IV}\sin\phi_{\text{pf}}$$

$$Q = |S|\sin\phi_{\text{pf}} = |I|^2 X = \frac{|V|^2}{|Z|^2}X = \frac{V^2}{X_C}$$

Rearrange and solve for the capacitive reactance.

$$\begin{aligned}Q &= \frac{V^2}{X_C} \\ X_C &= \frac{V^2}{Q} = \frac{(208\text{ V})^2}{-50\text{ VAR}} = -865\ \Omega\end{aligned}$$

Use the capacitive reactance to calculate the required capacitance.

Phasor Transforms of Sinusoids

$$\begin{aligned}X_C &= \frac{-1}{2\pi fC} \\ C &= \frac{-1}{2\pi f X_C} = \frac{-1}{2\pi(60\text{ Hz})(-865\ \Omega)} = 3.07\times 10^{-6}\text{ F}\quad (3\ \mu\text{F})\end{aligned}$$

The answer is (B).

11. The governing code is in *National Electrical Code* (NEC) Art. 210, Branch Circuits. According to NEC Sec. 210.21(B)(2) and Table 210.21(B)(2), a 20 A circuit is allowed a maximum of 16 A of cord- and plug-connected load. Therefore, only three of the control systems may be plugged into the receptacle, or any combination of the receptacles, for a total of 15 A.

The answer is (C).

12. A capacitor is being used that has a different voltage rating than the system. The reactive power can be found from a ratio of reactive power formulas. The reactive power for one phase of the system is

Phasor Transforms of Sinusoids

$$Q = IV\sin\phi_{\text{pf}}$$

$$Q = |S|\,\sin\phi_{\text{pf}} = |I|^2 X = \frac{|V|^2}{|Z|^2}X$$

$$= \frac{V^2}{X_C}$$

Consider a ratio of the reactive power at 208 V to the rated 440 V. The capacitive reactance is the same at each voltage because capacitive reactance depends on frequency and capacitance only. In other words, the reactive power varies with the square of the voltage.

$$\frac{Q_{208\text{ V}}}{Q_{440\text{ V}}} = \frac{\dfrac{V_{208\text{ V}}^2}{X_{208\text{ V}}}}{\dfrac{V_{440\text{ V}}^2}{X_{440\text{ V}}}}$$

$$= \frac{V_{208\text{ V}}^2}{V_{440\text{ V}}^2}$$

$$Q_{208\text{ V}} = Q_{440\text{ V}}\left(\frac{V_{208\text{ V}}^2}{V_{440\text{ V}}^2}\right)$$

$$= Q_{440\text{ V}}\left(\frac{V_{208\text{ V}}}{V_{440\text{ V}}}\right)^2$$

$$= (150\text{ kVAR})\left(\frac{208\text{ V}}{440\text{ V}}\right)^2$$

$$= 33.5\text{ kVAR}\quad(35\text{ kVAR})$$

The answer is (A).

13. The vector orthogonal to the electric and magnetic field strengths (the Poynting vector), **S**, is found from the cross product of **E** and **H**.

$$\mathbf{E}\times\mathbf{H} = \begin{vmatrix} \mathbf{i} & E_x & H_x \\ \mathbf{j} & E_y & H_y \\ \mathbf{k} & E_z & H_z \end{vmatrix}$$

$$= \begin{vmatrix} \mathbf{i} & 3 & 2 \\ \mathbf{j} & 7 & -3 \\ \mathbf{k} & 3 & 1 \end{vmatrix}$$

$$= \mathbf{i}\begin{vmatrix} 7 & -3 \\ 3 & 1 \end{vmatrix} - 3\begin{vmatrix} \mathbf{j} & -3 \\ \mathbf{k} & 1 \end{vmatrix} + 2\begin{vmatrix} \mathbf{j} & 7 \\ \mathbf{k} & 3 \end{vmatrix}$$

$$= \mathbf{i}\big(7-(-9)\big) - (3)\big(\mathbf{j}-(-3\mathbf{k})\big) + (2)(3\mathbf{j}-7\mathbf{k})$$

$$= 16\mathbf{i} + 3\mathbf{j} - 23\mathbf{k}\quad[\text{in W/m}^2]$$

If the Poynting vector had not been found (i.e., if $\mathbf{H}\times\mathbf{E}$ had been determined merely to find an orthogonal vector), that vector would have been $-16\mathbf{j} - 3\mathbf{j} + 23\mathbf{k}$. Such a vector is not one of the possible solutions.

The answer is (D).

14. Unbalanced current flow would remain unchanged by the resistor. Further, for an ungrounded system, normally no neutral is available. Therefore, option A and option D are not possible purposes.

Resistors in ground connections can be used to limit current during fault conditions, but such a connection differs from that shown. Therefore, option C is not the purpose.

An ungrounded system is never truly ungrounded. The coupling capacitors, C_0, provide a path to ground. As the ungrounded system expands, the size of the total capacitance becomes large enough for the capacitive ground fault current to become self-sustaining. However, the total capacitance is still low enough that it will not clear. Therefore, a resistor is added to the neutral connection to ensure sufficient current flow so protective devices can detect the fault and open circuit breakers (not shown).

The answer is (B).

15. The voltmeter in terms of full-scale values is shown.

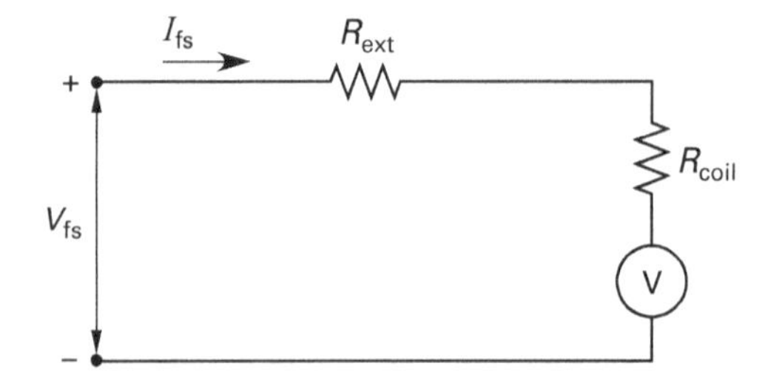

The sensitivity is 1 divided by the full-scale current. Using Ohm's law,

$$\begin{aligned} V_{fs} &= I_{fs}(R_{ext} + R_{coil}) \\ \frac{1}{I_{fs}} &= \frac{R_{ext} + R_{coil}}{V_{fs}} \\ &= 1000\ \Omega/\text{V} \end{aligned}$$

Rearrange and substitute the given values to determine the external resistance required.

$$\begin{aligned} R_{ext} &= V_{fs}\left(1000\ \frac{\Omega}{\text{V}}\right) - R_{coil} \\ &= (10\ \text{V})\left(1000\ \frac{\Omega}{\text{V}}\right) - 250\ \Omega \\ &= 9750\ \Omega \end{aligned}$$

Since an accuracy of 1% is specified for the voltmeter, an additional 1.5% error can occur in the external resistance, for a total accuracy of 2.5%. Therefore,

$$\begin{aligned} R_{ext,max} &= 9750\ \Omega + (0.015)(9750\ \Omega) \\ &= 9896.25\ \Omega \quad (9850\ \Omega) \end{aligned}$$

The value 9850 Ω was selected since the 9900 Ω value (while it does represent the rounded value of the answer) would exceed the circuit specifications for accuracy.

The answer is (B).

16. The illuminance, E, is given by luminous flux, Φ, divided by the area of concern, which in this case is the area of the hemispheric lamp.

$$\begin{aligned} E &= \frac{\Phi}{A} = \frac{\Phi}{\frac{1}{2}A_{sphere}} \\ &= \frac{3000\ \text{lm}}{\left(\frac{1}{2}\right)(4\pi)(5\ \text{m})^2} \\ &= 19.1\ \text{lx} \quad (20\ \text{lx}) \end{aligned}$$

The answer is (A).

17. Rung numbers may not always be visible. Remember that rungs are numbered from the top of a coding schematic to the bottom. Additionally, all the items attached to a given line are considered to be in the same rung. For example, in rung 1, the START and STOP switch, CR1, and the CR1 maintaining contact around the START push-button are all part of the same rung.

Rung 3 is the rung consisting of CR2 normally open contact and relay PL2. Therefore, the rule for rung 3 is that CR2 must be energized.

However, for CR2 to be energized, the START button must be pressed, and CR1 must be energized, which then energizes the timer. The timer sets the enable (EN) line ON or high, which energizes CR2, and closes the CR2 normally open contact in rung 3, turning PL2 ON.

PL2 is thus a relay that reflects the status of the timer enable line. This statement could also be listed as the rule for rung 3, though the rules normally focus on the rung itself.

Putting all of the conditions together gives the most complete answer.

The answer is (D).

18. When a ground occurs on a phase of an ungrounded system, the reference point shifts from the neutral to the grounded phase. The following illustration shows the phase relationship.

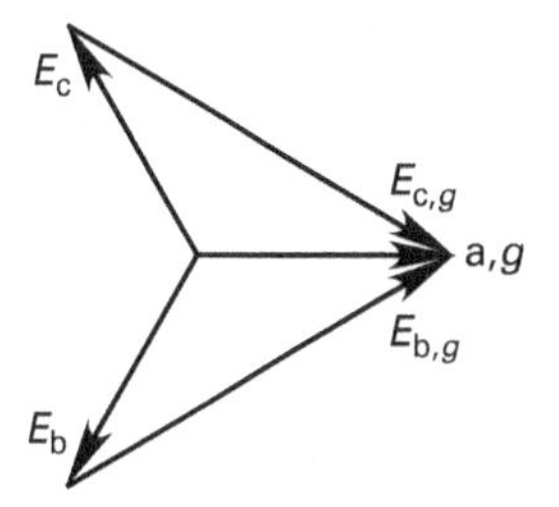

The illustration shows that the voltage between phase B and ground, $E_{b,g}$, becomes the phase sum of phases A and B. The addition of two phase values of equal magnitude but 120° apart results in a phasor $\sqrt{3}$ larger than and 30° offset from the original. The line voltage is

$$\begin{aligned} V_{b,g} &= \sqrt{3}\,(208\ \text{V}) \\ &= 360.3\ \text{V} \quad (360\ \text{V}) \end{aligned}$$

The answer is (D).

19. The resistance that results in a 3.0 V drop is calculated from Ohm's law.

$$\begin{aligned} V &= IR_{max} \\ R_{max} &= \frac{V}{I} = \frac{3.0\ \text{V}}{30\ \text{A}} \\ &= 0.10\ \Omega \end{aligned}$$

The resistance per unit length, R_ℓ, of the copper wire at 150°F (65.6°C) is

$$\begin{aligned} R_\ell &= R_{0,\ell}(1 + \alpha\Delta T) \\ &= \left(\frac{1\ \Omega}{1000\ \text{ft}}\right)\left(1 + (0.00402°\text{C}^{-1})(65.6°\text{C} - 20°\text{C})\right) \\ &= 1.18 \times 10^{-3}\ \Omega/\text{ft} \end{aligned}$$

At the operating temperature, the maximum length of wire that can be used, in meters, is

$$L = \frac{R_{\max}}{R_\ell} = \left(\frac{0.10\ \Omega}{1.18\times 10^{-3}\ \frac{\Omega}{\text{ft}}}\right)\left(0.305\ \frac{\text{m}}{\text{ft}}\right)$$
$$= 25.8\ \text{m}$$

The system is a two-wire DC system. The total length of the wire must be one-half of the length calculated.

$$L_{\text{DC}} = \frac{L}{2} = \frac{25.8\ \text{m}}{2}$$
$$= 12.9\ \text{m}\quad (13\ \text{m})$$

The answer is (C).

20. The equation for converting from one per-unit impedance value to another in a different base with the same base voltage is

Per Unit System

$$Z_{\text{pu,new}} = Z_{\text{pu,old}}\left(\frac{V_{\text{base,old}}}{V_{\text{base,new}}}\right)^2\left(\frac{S_{\text{base,new}}}{S_{\text{base,old}}}\right)$$
$$= (0.035)(1)^2\left(\frac{100\ \text{MVA}}{50\ \text{MVA}}\right)$$
$$= 0.07\quad (7\%)$$

The answer is (C).

21. The applicable formula for the rms (effective) secondary voltage is

Single-Phase Transformer Equivalent Circuits

$$v_s(t) = 4.44 N_s f \Phi_m \cos\omega t$$

Only the peak flux is desired. Therefore,

$$V_s = 4.44 N_s f \Phi_m$$
$$\Phi_m = \frac{V_s}{4.44 N_s f}$$

The rms secondary voltage is

Single-Phase Transformer Equivalent Circuits

$$\frac{V_s}{V_p} = \frac{N_s}{N_p}$$

$$V_s = V_p\left(\frac{N_s}{N_p}\right) = (110\ \text{V})\left(\frac{1}{10}\right)$$
$$= 11\ \text{V}$$

Substitute the rms secondary voltage into the formula for magnetic flux.

$$\Phi_m = \frac{V_s}{4.44 N_s f}$$
$$= \frac{11\ \text{V}}{(4.44)(1)(60\ \text{Hz})}$$
$$= 41.3\times 10^{-3}\ \text{Wb}\quad (40\ \text{mWb})$$

The same flux is present in the primary.

The answer is (C).

22. The transformers, though rated for the same line voltages, cannot be operated in parallel because the line voltages are out of phase by 30°. Option I is false.

A delta connection can suffer an open phase and still supply a three-phase load and maintain the same line voltages. Option II is true.

The wye connection, if ungrounded at the neutral, will suffer an increase in line voltage during a ground fault. The increase in voltage would be $\sqrt{3}$ of the original value. Option III is true.

The answer is (A).

23. The bulk of the *National Electrical Code* (NEC) and branch circuit requirements (i.e., the most used requirements) are contained in NEC Art. 210, Branch Circuits. Feeder requirements are in Art. 215. Article 220 contains guidance for load calculations. For general code questions, start in Art. 210 to determine the number of circuits required, then use Art. 220 to determine how to calculate the load and apply the necessary demand factors. Examples of calculations are given in Annex D of the NEC.

The articles referred to in the annex of the code are sometimes starting points that do not cover all relevant information. If the code requirement is not one that a "minimally competent engineer," per the evaluation of the NCEES, should know, then the code contents will be in the problem statement rather than in the solution, as shown here. General lighting loads can be placed on 15 A breakers and are therefore not included in this calculation. However, the loads must be calculated in accordance with Sec. 220.12.

The 20 A branch circuits required in a dwelling unit are determined in NEC Sec. 210.11(C). Two are required for small appliances (Sec. 210.11(C)(1)). One is required for laundry (Sec. 210.11(C)(2)). One is required for the bathroom (Sec. 210.11(C)(3)). Therefore, at least four branch circuits are required.

The answer is (B).

24. The torque of the motor at full load is

Synchronous Machines: Power, Torque, and Speed Relationships

$$P_{kW} = \frac{T_{N\cdot m} n_{rev/min}}{9549}$$

$$T_{N\cdot m} = \frac{9549 P_{kW}}{n_{rev/min}} = \frac{(9549)(373 \text{ kW})}{600 \ \frac{\text{rev}}{\text{min}}}$$
$$= 5936 \text{ N·m}$$

The starting torque is 125% of the motor's full-load torque.

$$T_{N\cdot m} = (1.25)(5936 \text{ N·m})$$
$$= 7420 \text{ N·m} \quad (7400 \text{ N·m})$$

The answer is (D).

25. The maximum current seen by the shunt during the discharge is 500 A. Therefore, the shunt must be rated for 500 A. By Ohm's law, the resistance is

$$V = IR$$
$$R = \frac{V}{I}$$
$$= \frac{50 \times 10^{-3} \text{ V}}{500 \text{ A}}$$
$$= 0.1 \times 10^{-3} \ \Omega \quad (100 \ \mu\Omega)$$

The answer is (B).

26. The correct voltage for charging a battery depends on the battery's chemistry (e.g., lead acid) and temperature. The age of the battery has no impact on its charging voltage.

The answer is (D).

27. When calculating a net load or a computed load (the overall load for a dwelling, including such items as the range, dryer, and other items not accounted for in the basic net load), one is determining the feeder and/or service sizing requirements. Start with *National Electrical Code* (NEC) Art. 220, Part III, and work through all applicable items. Each portion of Art. 220, Part III normally references earlier articles with the applicable requirements. (Article 220, Part IV contains the optional calculations.)

NEC Sec. 220.40 mentions general lighting demand factors. Article 220, Part II contains the guidance for calculating branch circuit loads, including general lighting. Section 220.12 specifies that general lighting is calculated using the square footage, not including unused or unfinished spaces. From NEC Table 220.12, the unit load is 3 VA/ft^2. Therefore,

$$P_{lighting} = (\text{unit load})(\text{area})$$
$$= \left(3 \ \frac{\text{VA}}{\text{ft}^2}\right)(2000 \text{ ft}^2)$$
$$= 6000 \text{ VA}$$

The variable P is used when S (for apparent power) would seem to be more applicable. This is done to emphasize that, in the code, the kW ratings of equipment are generally considered to be equivalent to the kVA ratings. See NEC Sec. 220.54 as an example. Additionally, the term "power" is routinely used. Nevertheless, circuitry must be designed to handle the apparent power.

Although a branch circuit is required for the bathroom, no additional load calculation is required since the bathroom's load is included in the general lighting calculation [NEC Sec. 220.14(J)(1)].

Do not yet apply the demand factor for general lighting.

Continuing in NEC Sec. 220, Part III, the next applicable item is "Small Appliances and Laundry Loads" in Sec. 220.52. Section 220.52(A) requires a load of 1500 VA for each two-wire small-appliance circuit required by Sec. 210.11(C)(1), which calls for two such circuits. Therefore, the small appliance load is

$$P_{small\ appliance} = \begin{pmatrix} \text{number of small} \\ \text{appliance circuits} \end{pmatrix}\left(1500 \ \frac{\text{VA}}{\text{circuit}}\right)$$
$$= (2 \text{ circuits})\left(1500 \ \frac{\text{VA}}{\text{circuit}}\right)$$
$$= 3000 \text{ VA}$$

Continuing in NEC Art. 220, Part III, the next applicable item is the "Laundry Circuit Load" in Sec. 220.52(B). This section requires 1500 VA for each two-wire laundry branch circuit required by Sec. 210.11(C)(2), which calls for only one such circuit. Therefore, the laundry load is

$$P_{laundry} = (\text{laundry load})\left(1500 \ \frac{\text{VA}}{\text{circuit}}\right)$$
$$= (1 \text{ circuit})\left(1500 \ \frac{\text{VA}}{\text{circuit}}\right)$$
$$= 1500 \text{ VA}$$

Per NEC Sec. 220.52(A) and (B), the small-appliance load and the laundry load may be included with the general lighting load and the demand factors of Sec. 220.42

for general lighting applied. The total "general lighting load" is

$$\begin{aligned} P_{\text{general}} &= P_{\text{lighting}} + P_{\text{small appliance}} + P_{\text{laundry}} \\ &= 6000\ \text{VA} + 3000\ \text{VA} + 1500\ \text{VA} \\ &= 10\,500\ \text{VA} \end{aligned}$$

Per NEC Table 220.42, for dwelling units, the first 3000 VA has a demand factor of 100%. The remaining 7500 VA of load (which ranges from 3000 VA to 10 500 VA) has a demand factor of 35%. Therefore, the net load with demand factors applied is

$$\begin{aligned} P_{\text{net}} &= (\text{first } 3000\ \text{VA})(1.0) \\ &\quad +(\text{next } 3000\ \text{VA to } 10\,500\ \text{VA})(0.35) \\ &= (3000\ \text{VA})(1.0) + (7500\ \text{VA})(0.35) \\ &= 5625\ \text{VA} \quad (5600\ \text{VA}) \end{aligned}$$

The answer is (A).

28. The speed regulation, SR, of the motor is

Governor Control for Synchronous Generators

$$\begin{aligned} \text{SR} &= \frac{n_{\text{nl}} - n_{\text{fl}}}{n_{\text{fl}}} \times 100\% \\ &= \frac{1190\ \text{rpm} - 1150\ \text{rpm}}{1150\ \text{rpm}} \times 100\% \\ &= 3.48\% \quad (3.5\%) \end{aligned}$$

The answer is (B).

29. Per *National Electrical Code* (NEC) Sec. 220.54, the minimum size that can be used for a dryer is 5 kW. Given that the dryer specified is greater than 5 kW, the nameplate rating of 6 kW should be used. Per the section, the kW ratings can be considered equivalent to kVA ratings. The demand factor for a single dryer is 100% per NEC Table 220.54. Therefore, the dryer load is 6000 VA.

Per NEC Sec. 220.55, the 8 kW is equivalent to 8 kVA. The demand, per NEC Table 220.55, Column B, is 8 kW, or 80%. Therefore, the range load is

$$\begin{aligned} P_{\text{range}} &= (8\ \text{kVA})(0.8)\left(1000\ \frac{\text{VA}}{\text{kVA}}\right) \\ &= 6400\ \text{VA} \end{aligned}$$

The net computed load, P_{ncl}, for the house is

$$\begin{aligned} P_{\text{ncl}} &= P_{\text{general}} + P_{\text{dryer}} + P_{\text{range}} + P_{\text{other}} \\ &= 4000\ \text{VA} + 6000\ \text{VA} + 6400\ \text{VA} + 0\ \text{VA} \\ &= 16\,400\ \text{VA} \end{aligned}$$

The net computed load current is

$$\begin{aligned} I_{\text{ncl}} &= \frac{P_{\text{ncl}}}{V} = \frac{16\,400\ \text{VA}}{240\ \text{V}} \\ &= 68.3\ \text{A} \end{aligned}$$

An additional consideration now comes into play. NEC Sec. 230.79(C) requires a one-family dwelling unit to have a minimum 100 A rating for the disconnecting means. Therefore, the minimum load is 100 A (see Sec. 230.79).

Using this 100 A minimum, one would normally now use the many tables of NEC Sec. 310, from Table 310.15 onward, to determine the conductor size. Again, an additional consideration comes into play. For single-phase dwelling services and feeders of 120/240 V, three-wire, single-phase type, NEC Sec. 310.15(B)(7) may be used. (This section allows higher current loads for a feeder of a given size. Or, put another way, it allows reduced feeder size. Note that the feeder size does not, in any case, need to be larger than the service conductor, per Sec. 215.2(A)(3).)

From NEC Sec. 310.15(B)(7)(1), using a minimum of 83% of 100 A, the copper wire used must be AWG 4, per NEC Table 310.15(B)(16).

The answer is (B).

30. The real power is

$$P = \sqrt{3}\, IE\text{pf}$$

The current is unknown in this equation. However, calculating the current is not required to find the real power output. The apparent power is

3-Phase Circuits

$$S = \sqrt{3}\, IE$$

Substitute S into the real power equation. The real power output is

$$\begin{aligned} P &= S\text{pf} \\ &= (15 \times 10^6\ \text{VA})(0.80) \\ &= 12 \times 10^6\ \text{W} \quad (12\ \text{MW}) \end{aligned}$$

The answer is (A).

31. *National Electrical Code* (NEC) Art. 240 contains the requirements for overload protection and lists the articles regarding specific equipment (see NEC Table 240.3). Article 240 should be the starting point for overload matters. (Overload requirements are not the same as short-circuit/ground-fault protection requirements.) Motor overload requirements are in Art. 430.

NEC Sec. 430.6(A)(1) indicates that the various tables given should be used to determine the current rating of the overload. NEC Table 430.248 is applicable and lists the current as 13.8 A. This is normally the current used to set the overload rating. However, in Sec. 430.6(A)(1), Exception 3 notes that the full-load current marked on the nameplate should be used, rather than the current taken from the tables, when both the current rating and horsepower rating are listed. (This is because the current rating is considered to be the more accurate of the two.) Therefore, 16 A is used as the full-load current.

Overload protection for motors is discussed in NEC Art. 430, Part III. From Sec. 430.32(A)(1), for a service factor not less than 1.15, the overload should be set at 125% of the full-load current rating. (For a service factor less than 1.15, the overload would be set at 115% of the full-load current rating.) Therefore,

$$\begin{aligned}\text{OL} &= I_{\text{fl}}(\text{OL factor}) \\ &= (16\ \text{A})(1.25) \\ &= 20\ \text{A}\end{aligned}$$

The answer is (C).

32. In the illustration, A is the locked-rotor torque, B is the pull-up torque, C is the pull-out torque, and D is the full-load torque.

The answer is (D).

33. Consider the items in the order given. According to *National Electrical Code* (NEC) Sec. 210.19(A)(1), the ampacity of the conductor must be equal to or greater than 100% of the noncontinuous load plus 125% of the continuous load before the application of derating factors.

$$\begin{aligned}\text{required ampacity} &= L_{\text{noncontinuous}} + 1.25L_{\text{continuous}} \\ &= 12\ \text{A} + (1.25)(10\ \text{A}) \\ &= 24.5\ \text{A}\end{aligned}$$

The THHN, 90°C column of NEC Table 310.15(B)(16) cannot be used for determining the final ampacity of the conductors because the terminals mentioned are rated for 60°C. (See Sec. 110.14(C)(1)(a)(2) for restrictions that are in place for termination provisions and conductors on circuits rated for 100 A or less, which are the majority of circuits under consideration.) The THHN, 90°C column of NEC Table 310.15(B)(16) can be used for "ampacity adjustment, correction, or both," according to Sec. 110.14(C). Therefore, using the 90°C column, the AWG 10 wire rated for 40 A is selected as the initial choice. (This is only the initial choice since more derating is required. The selection process is often iterative.)

The conductor rating must be adjusted for the 50°C ambient temperature. Using the correction factor in NEC Table 310.15(B)(2)(a), in the 90°C Temperature Rating of Conductor column and the 50°C ambient temperature row, the correction factor is 0.82. The design will also include five conductors in a given raceway. Per NEC Table 310.15(B)(3)(a), the adjustment factor is 80% for four to six conductors in the same raceway. Applying these two derating factors to the AWG 10 conductor ampacity gives

$$\begin{aligned}\begin{pmatrix}\text{adjusted} \\ \text{ampacity}\end{pmatrix} &= \begin{pmatrix}\text{rated} \\ \text{ampacity}\end{pmatrix}\begin{pmatrix}\text{temperature} \\ \text{derating}\end{pmatrix} \\ &\quad \times \begin{pmatrix}\text{conductors per} \\ \text{raceway derating}\end{pmatrix} \\ &\quad \times \begin{pmatrix}\text{other derating} \\ \text{factors}\end{pmatrix} \\ &= (40\ \text{A})(0.82)(0.80) \\ &= 26.2\ \text{A}\end{aligned}$$

The ampacity is greater than that required for the load, so the "conductors" are sized correctly with all derating accounted for, including temperature and multiple conductor adjustments.

To ensure that the rating of the 60°C terminal device is not exceeded, the determined ampacity of 26.2 A is compared to the ampacity allowed. From NEC Table 310.15(B)(16), the allowable ampacity is 30 A for an AWG 10 conductor and a temperature of 60°C. Therefore, the requirements of Sec. 110.14(C)(1)(a)(2) are met by the AWG 10 conductor.

A check of the AWG 12 conductor indicates that this conductor would be too small once derated to carry the expected load. (It would meet the 60°C terminal restriction since an AWG 12 conductor is rated for 25 A.)

A size AWG 10 conductor should be used.

The answer is (C).

34. The AC distance is

$$\begin{aligned}\text{AB}^2 + \text{BC}^2 &= \text{AC}^2 \\ \text{AC} &= \sqrt{\text{AB}^2 + \text{BC}^2} \\ &= \sqrt{(6\ \text{ft})^2 + (6\ \text{ft})^2} \\ &= 8.48\ \text{ft}\end{aligned}$$

The geometric mean distance, D_e, is

Line Inductance and Inductive Reactance

$$\begin{aligned}D_e &= \sqrt[3]{D_{\text{AB}}D_{\text{BC}}D_{\text{CA}}} \\ &= \sqrt[3]{(6\ \text{ft})(6\ \text{ft})(8.48\ \text{ft})} \\ &= 6.73\ \text{ft} \quad (7\ \text{ft})\end{aligned}$$

The answer is (B).

35. Ampacity of conductors is determined by the horsepower rating in the tables referenced in *National Electrical Code* (NEC) Sec. 430.6(A)(1) rather than by the nameplate ampere rating. NEC Table 430.250 is applicable to a three-phase squirrel-cage motor. Using the 5 hp and 230 V ratings as the input parameters, the full-load current is 15.2 A.

The ampacity must be 125% of the full-load rating per Sec. 430.22. Therefore,

$$\begin{aligned}\text{required ampacity} &= 1.25I_{\text{fl}}\\ &= (1.25)(15.2\ \text{A})\\ &= 19.0\ \text{A}\end{aligned}$$

The answer is (D).

36. Given that the transformer values are the base values, the base power is the rated power of transformer T_1, 150 MVA. The base voltage is the primary voltage of the transformer. In this problem, the primary voltage is the generator's voltage, 23 kV.

The base impedance is

$$\begin{aligned}Z_{\text{base}} &= \frac{V_{\text{base}}^2}{S_{\text{base}}} = \frac{(23\times 10^3\ \text{V})^2}{150\times 10^6\ \text{VA}}\\ &= 3.53\ \Omega \quad (3.5\ \Omega)\end{aligned}$$

The answer is (B).

37. *National Electrical Code* (NEC) Sec. 430.6(A)(2) requires the use of the nameplate current rating when determining the overload protection setting. A separate overload device, rather than an integral overload device, for a motor greater than 1 hp is covered in Sec. 430.32(A)(1). With a service factor less than 1.15, the motor overload should be set for 115% of the motor nameplate current rating. Therefore,

$$\begin{aligned}\text{OL setting} &= (\text{nameplate rating})(\text{adjustment factor})\\ &= (14\ \text{A})(1.15)\\ &= 16.1\ \text{A}\end{aligned}$$

The answer is (B).

38. The resistance of the generator is generally minimal, allowing the generator to be shown with a value of reactance only, as opposed to a value with resistance and reactance, on a one-line diagram. Option A is true.

The per-unit impedance of all transformers is the same on both sides, which allows distribution systems to be shown as one-line diagrams. Option B is true.

The power factor is 0.8 lagging. The impedance angle is

$$\begin{aligned}\theta &= \arccos 0.8\\ &= 37^\circ\end{aligned}$$

Since the power factor is lagging, the current angle (with the voltage considered as the reference at 0°) is −37°. This load phase relationship is reflected to the generator since using two transformers shifts the angle first one direction and then the other. Option C is true.

In any step-up delta-to-wye transformer, the positive sequence voltages and currents on the higher-voltage side lead the corresponding values on the lower-voltage side by 30°. (Consider the addition of any two wye-side phase value phasors at 120° apart resulting in a line phasor 30° ahead of its associated phase.) The transmission line (which is the higher-voltage side of T_1) has a current angle that is actually 30° ahead of the current angle of the generator (which matches the load values in this case). Therefore, the transmission line angle is −7° (30° ahead of −37°). Option D is false.

The situation is shown in the following illustration.

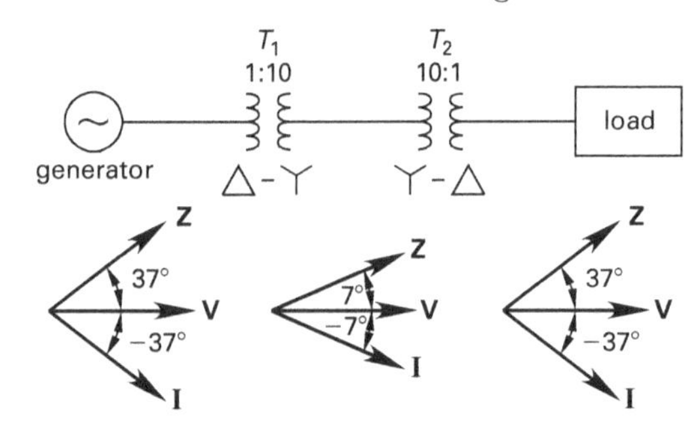

The answer is (D).

39. Information on short-circuit and ground fault protection for motor branch circuits is in *National Electrical Code* (NEC) Sec. 430.52, which redirects to NEC Table 430.52. From the table for "other than design B" motors, the rating should be 300% of the full-load current. From NEC Table 430.250, I_{fl} is 15.2 A.

$$\begin{aligned}\text{fuse rating} &= I_{\text{fl}}(3.00) = (15.2\ \text{A})(3.00)\\ &= 45.6\ \text{A}\end{aligned}$$

Since 45.6 A is a nonstandard size, the next higher size can be used, per NEC Sec. 430.52(C)(1), Exception 1. Standard-size fuses are given in Sec. 240.6(A). The next available size is 50 A. (This 50 A size does not exceed the 400% limit over the full-load current set by Sec. 430.52(C)(1), Exception 2(a).)

The answer is (D).

40. Rearrange the apparent power equation to find the line current.

3-Phase Circuits

$$S = IV$$

$$\begin{aligned}I &= \frac{S}{V} = \frac{300\times 10^3\ \text{VA}}{11\times 10^3\ \text{V}}\\ &= 27.3\ \text{A}\end{aligned}$$

The answer is (B).

Solutions
Exam 1: Afternoon Session

Content in blue refers to the *NCEES Handbook*.
Content in red is additional essential information.

41. The determinant is found using the cofactor expansion method.

$$\begin{aligned} A &= \begin{vmatrix} a & b & c \\ d & e & f \\ g & h & i \end{vmatrix} \\ &= a\begin{vmatrix} e & f \\ h & i \end{vmatrix} - d\begin{vmatrix} b & c \\ h & i \end{vmatrix} + g\begin{vmatrix} b & c \\ e & f \end{vmatrix} \\ &= \begin{vmatrix} 3 & 4 & 2 \\ 1 & 6 & 5 \\ 4 & 3 & 2 \end{vmatrix} \\ &= 3\begin{vmatrix} 6 & 5 \\ 3 & 2 \end{vmatrix} - 4\begin{vmatrix} 1 & 5 \\ 4 & 2 \end{vmatrix} + 2\begin{vmatrix} 1 & 6 \\ 4 & 3 \end{vmatrix} \\ &= (3)(12-15) - (4)(2-20) + (2)(3-24) \\ &= 21 \end{aligned}$$

The answer is (D).

42. Using Kirchhoff's voltage laws, the source voltage is

$$\begin{aligned} \mathbf{V}_{\text{source}} &= \mathbf{I}_{\text{line}}\mathbf{Z}_{\text{line}} + \mathbf{V}_{\text{load}} + \mathbf{I}_{\text{line}}\mathbf{Z}_{\text{line}} \\ &= 2\mathbf{I}_{\text{line}}\mathbf{Z}_{\text{line}} + \mathbf{V}_{\text{load}} \\ &= (2)\left(\frac{P}{V_{\text{terminal}}(\text{pf})}\angle 0^\circ\right)\mathbf{Z}_{\text{per 1000 ft}}\left(\frac{180 \text{ ft}}{1000 \text{ ft}}\right) \\ &\quad + \mathbf{V}_{\text{load}} \\ &= (2)\left(\frac{5000 \text{ W}}{(220 \text{ V})(1)}\angle 0^\circ\right)(0.8 + j0.5 \ \Omega)\left(\frac{180 \text{ ft}}{1000 \text{ ft}}\right) \\ &\quad + 220 \text{ V}\angle 0^\circ \\ &= (36.36 + j22.73 \text{ V})\left(\frac{180 \text{ ft}}{1000 \text{ ft}}\right) + 220 \text{ V}\angle 0^\circ \\ &= (6.54 + j4.09 \text{ V}) + 220 \text{ V}\angle 0^\circ \\ &= 226.54 + j4.09 \text{ V} \\ &= 226.58 \text{ V}\angle 1.03^\circ \\ &= 227 \text{ V} \quad (230 \text{ V}) \end{aligned}$$

The answer is (B).

43. Projects of differing lengths can be compared using their present worth, P.

Engineering Economics

$$P = F(P/F, i\%, n)$$

The present worth of design A in millions is

$$\begin{aligned} P_{\text{A}} &= -\$200 + (\$350)(P/F, 5\%, 10) \\ &= -\$200 + (\$350)(0.6139) \\ &= \$14.86 \end{aligned}$$

The present worth of design B in millions is

$$\begin{aligned} P_{\text{B}} &= -\$60 + (\$80)(P/F, 5\%, 5) \\ &= -\$60 + (\$80)(0.7835) \\ &= \$2.68 \end{aligned}$$

The present worth of design C in millions is

$$\begin{aligned} P_{\text{C}} &= -\$20 + (\$30)(P/F, 5\%, 20) \\ &= -\$20 + (\$30)(0.3769) \\ &= -\$8.69 \end{aligned}$$

Design A has the greatest present worth and is the economically superior alternative.

The answer is (A).

44. The operating speed is calculated from the following equation for slip.

Percent Slip in Induction Machines

$$\begin{aligned} s &= \frac{n_s - n}{n_s} \\ n &= n_s - sn_s \\ &= n_s(1 - s) \end{aligned}$$

The synchronous speed is

Synchronous Machines: Synchronous Speed

$$\begin{aligned} n_s &= \frac{120f}{p} = \frac{(120)(60 \text{ Hz})}{2} \\ &= 3600 \text{ rpm} \end{aligned}$$

Substitute this synchronous speed into the slip equation.

$$\begin{aligned} n &= n_s(1-s) \\ &= (3600 \text{ rpm})(1-0.03) \\ &= 3492 \text{ rpm} \quad (3500 \text{ rpm}) \end{aligned}$$

The answer is (C).

45. The impedance is measured at the same frequency at which the circuit operates. The frequency of the circuit is

$$377 \ \frac{\text{rad}}{\text{s}} = 2\pi f$$

$$\begin{aligned} f &= \frac{377 \ \frac{\text{rad}}{\text{s}}}{2\pi} \\ &= 60 \text{ 1/s} \quad (60 \text{ Hz}) \end{aligned}$$

The total impedance of L_1 is 40 mH plus the mutual inductance.

$$\begin{aligned} L_{1,\text{total}} &= L_1 + L_{\text{mutual}} = 40 \text{ mH} + 15 \text{ mH} \\ &= 55 \text{ mH} \end{aligned}$$

The situation is as shown.

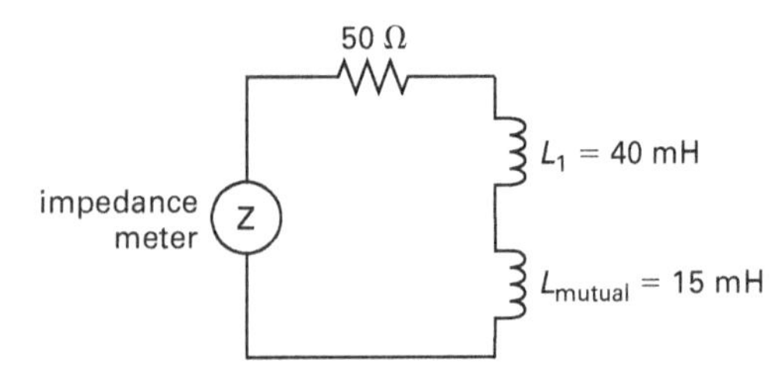

The total primary impedance (measured by the impedance instrument) is

$$\begin{aligned} Z &= R + L_1 + L_{\text{mutual}} \\ &= 50 \ \Omega + j\omega L_1 + j\omega L_{\text{mutual}} \\ &= 50 \ \Omega + j\left(377 \ \frac{\text{rad}}{\text{s}}\right)(40 \times 10^{-3} \text{ H}) \\ &\quad + j\left(377 \ \frac{\text{rad}}{\text{s}}\right)(15 \times 10^{-3} \text{ H}) \\ &= 50 \ \Omega + j15.08 \ \Omega + j5.66 \ \Omega \\ &= 50 + j20.74 \ \Omega \\ &= 54.13 \ \Omega\angle 22.53^\circ \quad (54 \ \Omega\angle 23^\circ) \end{aligned}$$

The answer is (C).

46. The real power of the motor is

$$\begin{aligned} P &= \sqrt{3} I_l V_l (\text{pf}) = \sqrt{3}(17 \text{ A})(220 \text{ V})(0.8) \\ &= 5182 \text{ W} \end{aligned}$$

The reactive power of the motor operating at a power factor of 0.8 is

Complex Power Triangle (Inductive Load)

$$\begin{aligned} Q_{0.8} &= P \tan(\arccos \text{pf}) \\ &= (5182 \text{ W})\tan(\arccos 0.8) \\ &= (5182 \text{ W})(0.75) \\ &= 3886 \text{ VAR} \end{aligned}$$

The reactive power of the motor operating at a power factor of 0.9 is

$$\begin{aligned} Q_{0.9} &= P \tan(\arccos \text{pf}) \\ &= (5182 \text{ W})\tan(\arccos 0.9) \\ &= (5182 \text{ W})(0.48) \\ &= 2510 \text{ VAR} \end{aligned}$$

The reactive power required to change the power factor from 0.8 to 0.9 is

$$\begin{aligned} Q_{\text{correction}} &= Q_{0.9} - Q_{0.8} \\ &= 2510 \text{ VAR} - 3886 \text{ VAR} \\ &= -1376 \text{ VAR} \quad (-1380 \text{ VAR}) \end{aligned}$$

The answer is (C).

47. Reflect the secondary impedance to the primary.

Single-Phase Transformer Equivalent Circuits

$$\sqrt{\frac{Z_p}{Z_s}} = \frac{N_p}{N_s}$$

$$\frac{Z_p}{Z_s} = \left(\frac{N_p}{N_s}\right)^2$$

$$\begin{aligned} Z_p &= Z_s\left(\frac{N_p}{N_s}\right)^2 \\ &= (100 \times 10^3 \ \Omega)\left(\frac{33}{1}\right)^2 \\ &= 108.9 \times 10^6 \ \Omega \end{aligned}$$

The current in the primary winding is

$$\mathbf{V} = \mathbf{IZ}$$

Since the conditions for the transformer are ideal, the following relationship is also true.

$$V = IR$$

Substitute the given and calculated values, noting that Z_p is equal to R_p in this case.

$$\begin{aligned} V_p &= I_p R_p \\ I_p &= \frac{V_p}{R_p} \\ &= \frac{110 \text{ V}}{108.9 \times 10^6 \ \Omega} \\ &= 1.01 \times 10^{-6} \text{ A} \quad (1 \ \mu\text{A}) \end{aligned}$$

The answer is (A).

48. The power triangle illustrates the relationships between variables.

Complex Power Triangle (Inductive Load)

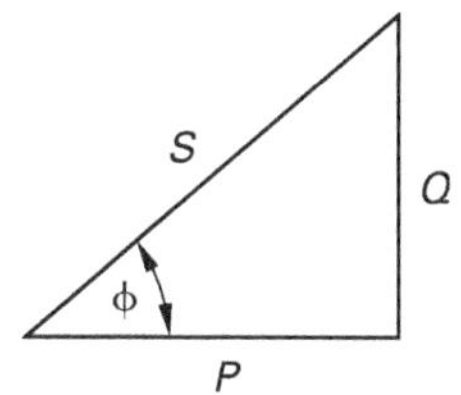

The initial reactive power is

$$\begin{aligned} S_0 &= \frac{P}{\text{pf}} = \frac{5000 \text{ kW}}{0.7} \\ &= 7143 \text{ kVA} \\ Q_0 &= \sqrt{S^2 - P^2} = \sqrt{(7143 \text{ kVA})^2 - (5000 \text{ kW})^2} \\ &= 5101 \text{ kVAR} \end{aligned}$$

The final reactive power is

$$\begin{aligned} S_f &= \frac{P}{\text{pf}} = \frac{5000 \text{ kW}}{0.8} \\ &= 6250 \text{ kVA} \\ Q_f &= \sqrt{S^2 - P^2} = \sqrt{(6250 \text{ kVA})^2 - (5000 \text{ kW})^2} \\ &= 3750 \text{ kVAR} \end{aligned}$$

The reactive power required in the capacitor bank is

$$\begin{aligned} Q_{\text{bank}} &= Q_f - Q_0 = 3750 \text{ kVAR} - 5101 \text{ kVAR} \\ &= -1351 \text{ kVAR} \\ |Q_{\text{bank}}| &= 1351 \text{ kVAR} \quad (1350 \text{ kVAR}) \end{aligned}$$

The answer is (A).

49. The total electricity, or power, used in a day is

$$\begin{aligned} P_{\text{kW·h/d}} &= P_{\text{kW}} t_{\text{h/d}} = (100 \text{ W})\left(\frac{1 \text{ kW}}{1000 \text{ W}}\right)\left(6 \ \frac{\text{h}}{\text{d}}\right) \\ &= 0.6 \text{ kW·h/d} \end{aligned}$$

The daily cost is

$$\begin{aligned} C_{\text{daily}} &= P_{\text{kW·h/d}} C_{\text{kW·h}} = \left(0.6 \ \frac{\text{kW·h}}{\text{d}}\right)\left(\frac{\$0.07}{\text{kW·h}}\right) \\ &= \$0.042/\text{d} \end{aligned}$$

The annual cost is

$$\begin{aligned} C_{\text{annual}} &= C_{\text{daily}}\left(365 \ \frac{\text{d}}{\text{yr}}\right) = \left(\frac{\$0.042}{\text{d}}\right)\left(365 \ \frac{\text{d}}{\text{yr}}\right) \\ &= \$15.33/\text{yr} \quad (\$15.00/\text{yr}) \end{aligned}$$

The answer is (D).

50. Calculate the capacitive reactance and impedance of the capacitor.

Phasor Transforms of Sinusoids

$$\begin{aligned} X_C &= \frac{1}{2\pi f C} = \frac{1}{2\pi(60 \text{ Hz})(60 \times 10^{-6} \text{ F})} \\ &= 44.21 \ \Omega \\ \mathbf{Z}_C &= X_C \angle -90^\circ \\ &= 44.2 \ \Omega \angle -90^\circ \end{aligned}$$

Calculate the impedance of the parallel combination of the capacitor and the 5 Ω resistor.

$$\begin{aligned} \mathbf{Z}_C || \mathbf{R}_{5\Omega} &= \frac{\mathbf{Z}_C \mathbf{R}_{5\Omega}}{\mathbf{Z}_C + \mathbf{R}_{5\Omega}} \\ &= \frac{(44.2 \ \Omega\angle -90^\circ)(5 \ \Omega\angle 0^\circ)}{44.2 \ \Omega\angle -90^\circ + 5 \ \Omega\angle 0^\circ} \\ &= 4.97 \ \Omega\angle -6^\circ \end{aligned}$$

Use the voltage divider equation to obtain the Thevenin equivalent voltage.

$$\begin{aligned} \mathbf{V}_{\text{Th}} &= \mathbf{V}_{\text{source}}\left(\frac{\mathbf{Z}_{||}}{\mathbf{Z}_{||} + \mathbf{Z}_{10}}\right) \\ &= (10 \text{ V}\angle 0^\circ)\left(\frac{4.97 \ \Omega\angle -6^\circ}{4.97 \ \Omega\angle -6^\circ + 10 \ \Omega\angle 0^\circ}\right) \\ &= (10 \text{ V}\angle 0^\circ)(0.333\angle -4^\circ) \\ &= 3.3 \text{ V}\angle -4^\circ \end{aligned}$$

The answer is (A).

51. Given two independent sets of events, a and b, such as the failure of generators A and B, the probability that events a_i and b_i will both occur is

Reliability

$$p\{a_i \text{ and } b_i\} = p\{a_i\}p\{b_i\}$$

Substituting the given probabilities results in the likelihood of a failure of both generators and the subsequent inability to meet peak demand.

$$p\{a_i\}p\{b_i\} = (0.02)(0.04) = 0.0008$$

The reliability, R, of the system is

$$R = 1 - p\{\text{failure}\}$$

The likelihood that peak demand will be met is

$$\begin{aligned} R &= 1 - p\{\text{failure}\} \\ &= 1 - 0.0008 \\ &= 0.9992 \end{aligned}$$

The answer is (D).

52. The neutral conductor is defined in *National Electrical Code* Art. 100 as the conductor connected to the neutral point of the system that is intended to carry current under normal conditions. The neutral point is the place where the vectorial sum on the nominal voltages from all other phases within the system that utilize the neutral, with respect to the neutral point, is zero potential.

The answer is (D).

53. Conductor B meets this definition. The terms associated with the other conductors are shown.

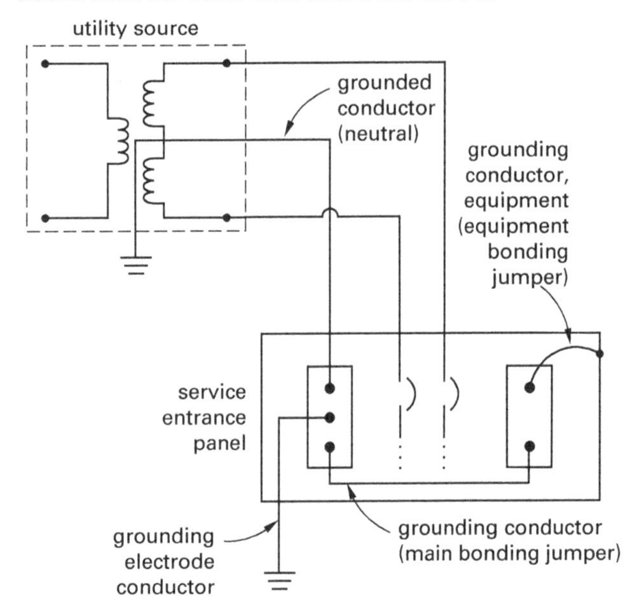

Definitions for each of the items are found in *National Electrical Code* (NEC) Art. 100. Information on grounding is found in NEC Art. 250.

The answer is (B).

54. The turns ratio is determined from the ratio of the primary to secondary phase voltages. The phase voltage on the primary side is

3-Phase Circuits

$$\begin{aligned} V_{\phi\,\text{primary}} &= \frac{V_\text{line}}{\sqrt{3}} = \frac{11 \times 10^3\ \text{V}}{\sqrt{3}} \\ &= 6351\ \text{V} \end{aligned}$$

Because of the delta connection, the phase voltage on the secondary is identical to the line voltage, 480 V. The turns ratio is

Single-Phase Transformer Equivalent Circuits

$$\begin{aligned} a &= \frac{n_\text{primary}}{n_\text{secondary}} = \frac{V_\text{primary}}{V_\text{secondary}} \\ &= \frac{6351\ \text{V}}{480\ \text{V}} \\ &= 13.23 \quad (13) \end{aligned}$$

The answer is (A).

55. Apply Kirchhoff's current law (KCL) at the inverting input.

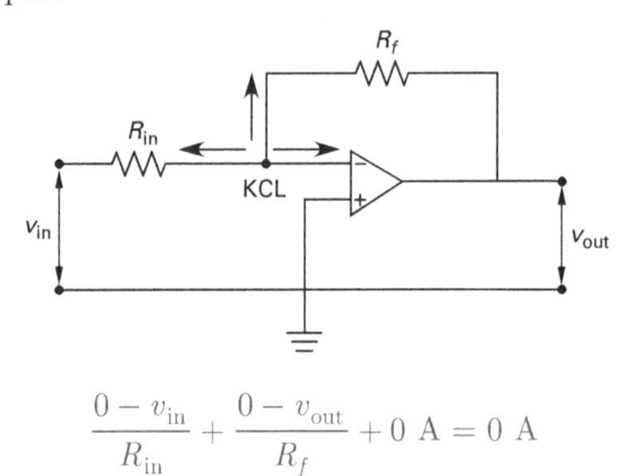

$$\frac{0 - v_\text{in}}{R_\text{in}} + \frac{0 - v_\text{out}}{R_f} + 0\ \text{A} = 0\ \text{A}$$

The voltage gain, A_v, is the output voltage divided by the input voltage. Rearrange to solve for the gain.

$$\begin{aligned} \frac{0 - v_\text{out}}{R_f} &= \frac{v_\text{in}}{R_\text{in}} \\ \frac{v_\text{out}}{v_\text{in}} &= -\frac{R_f}{R_\text{in}} \\ A_v &= \frac{v_\text{out}}{v_\text{in}} \\ &= -\frac{R_f}{R_\text{in}} \\ &= -\frac{1 \times 10^6\ \Omega}{25\ \Omega} \\ &= -40\,000 \end{aligned}$$

The answer is (A).

56. The number of poles is

Synchronous Machines: Synchronous Speed

$$n_s = \frac{120f}{p}$$

$$p = \frac{120f}{n_s}$$

The synchronous speed is not given but can be determined from the slip and nameplate speed.

Percent Slip in Induction Machines

$$s = \frac{n_s - n}{n_s}$$

$$\begin{aligned} n_s &= \frac{n}{1-s} \\ &= \frac{1620 \text{ rpm}}{1-0.10} \\ &= 1800 \text{ rpm} \end{aligned}$$

Substitute the synchronous speed into the equation for the number of poles.

$$\begin{aligned} p &= \frac{120f}{n_s} \\ &= \frac{(120)(60 \text{ Hz})}{1800 \text{ rpm}} \\ &= 4 \end{aligned}$$

The answer is (B).

57. The reflected impedance is

Single-Phase Transformer Equivalent Circuits

$$Z_{\text{ref}} = Z_{\text{ep}} = a^2 Z_s$$

$$\begin{aligned} &= \left(\frac{N_p}{N_s}\right)^2 Z_s \\ &= \left(\frac{3}{1}\right)^2 (3.0 \ \Omega\angle 30°) \\ &= 27 \ \Omega\angle 30° \end{aligned}$$

The answer is (C).

58. Per *National Electrical Safety Code* Part 1, Sec. 11, Art. 111, Table 111-1, the illumination level for control rooms, Type A (large centralized control room 1.68 m above the floor) is 270 lx.

The answer is (D).

59. A ground-fault circuit interrupter (GFCI) basic circuit is shown.

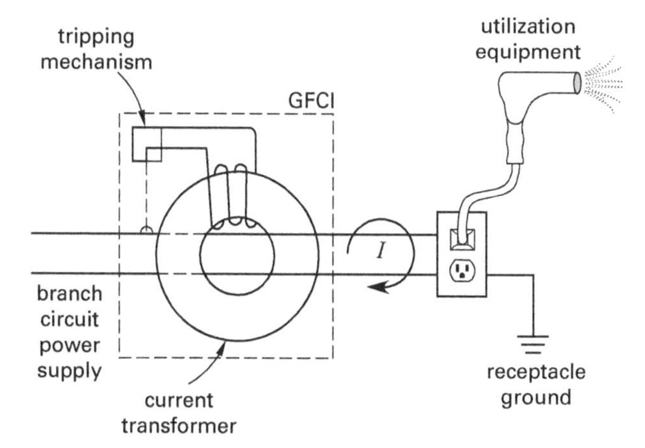

When the current is unbalanced due to a ground fault, the unbalanced magnetic fields induce a net voltage in the transformer, the current flows in the tripping mechanism, and the GFCI contact opens, protecting any person holding the offending equipment. Therefore, the conditions of options A, B, and C occur.

The answer is (D).

60. The voltage regulation is

Transformer's Percentage Voltage Regulation

$$\text{regulation} = \frac{V_{\text{nl}} - V_{\text{fl}}}{V_{\text{fl}}} \times 100\%$$

$$\begin{aligned} &= \frac{1.03 \text{ pu} - 1.00 \text{ pu}}{1.00 \text{ pu}} \times 100\% \\ &= 3.00\% \end{aligned}$$

The answer is (A).

61. The characteristic impedance of the transmission line is

Transmission Line Models

$$Z_0 = \sqrt{\frac{L_l}{C_l}} = \sqrt{\frac{500 \times 10^{-6} \ \frac{\text{H}}{\text{mi}}}{0.100 \times 10^{-6} \ \frac{\text{F}}{\text{mi}}}}$$

$$\begin{aligned} &= \sqrt{5.00 \times 10^3 \ \frac{\text{H}}{\text{F}}} \\ &= \sqrt{5.00 \times 10^3 \ \frac{\text{V}^2}{\text{A}^2}} \\ &= 70.7 \ \Omega \end{aligned}$$

The answer is (D).

62. The positive sequence current for phase A, noting that **a** is $1\angle 120°$, is

Symmetrical Components

$$\begin{aligned} I_A &= \tfrac{1}{3}(I_A + aI_B + a^2 I_C) \\ &= \left(\frac{1}{3}\right)(5\ \text{A}\angle 0° + (1\angle 120°)(5\ \text{A}\angle 180°) + 0) \\ &= \left(\frac{1}{3}\right)(5\ \text{A}\angle 0° + 5\ \text{A}\angle 300°) \\ &= \left(\frac{1}{3}\right)(8.66\ \text{A}\angle{-30°}) \\ &= 2.9\ \text{A}\angle{-30°} \quad (3\ \text{A}\angle{-30°}) \end{aligned}$$

The answer is (A).

63. The protective margin is

$$\begin{aligned} \text{PM} &= \frac{V_{\text{withstand}} - V_{\text{protection}}}{V_{\text{protection}}} \\ &= \frac{90\ \text{kV} - 38\ \text{kV}}{38\ \text{kV}} \\ &= 1.37 \quad (140\%) \end{aligned}$$

The answer is (D).

64. The current required to operate the relay is

$$\begin{aligned} I_{\text{CT}} &= aI_{\text{primary}} \\ &= \left(\frac{5}{400}\right)(160\ \text{A}) \\ &= 2.0\ \text{A} \end{aligned}$$

The answer is (A).

65. Phase quantities are given in the problem statement. Therefore, conversion from line quantities is not necessary. First, calculate the base impedance.

Per Unit System

$$\begin{aligned} Z_{\text{base}} &= \frac{V_p^2}{S_p} = \frac{(15\ \text{kV})^2}{30\ \text{kVA}} \\ &= 7.5\ \text{k}\Omega \quad (7500\ \Omega) \end{aligned}$$

The per-unit impedance is

$$\begin{aligned} Z_{\text{pu}} &= \frac{Z_{\text{actual}}}{Z_{\text{base}}} = \frac{75\ \Omega}{7500\ \Omega} \\ &= 0.01\ \text{pu} \end{aligned}$$

The answer is (A).

66. Nearly all fault studies assume that generated voltage is unaffected by machine speed variations, only synchronous currents and voltages are considered in stator windings and the power system (meaning DC offsets currents and harmonics are ignored), and symmetrical components are used to represent unbalanced faults.

Faults can occur at any point in the system; option A is not an assumption.

The answer is (A).

67. The current is

$$\begin{aligned} I_A + I_B + I_N &= 0\ \text{A} \\ |-I_N| &= |I_A + I_B| \end{aligned}$$

The current is determined from the power and the voltage. However, the voltage necessary in the calculation is the phase voltage, which is found from the phase-to-phase (line) voltage, as follows.

3-Phase Circuits

$$\begin{aligned} V_p &= \frac{V_{\text{pp}}}{\sqrt{3}} \\ &= \frac{12.8\ \text{kV}}{\sqrt{3}} \\ &= 7.39\ \text{kV} \end{aligned}$$

Using the calculated voltage and the given power provides the neutral current.

$$\begin{aligned} |-I_N| &= |I_A + I_B| \\ &= \left|\frac{S_A}{V_A} + \frac{S_B}{V_B}\right| \\ &= \left|\frac{150 + j75\ \text{kVA}}{7.39\ \text{kV}\angle 0°} + \frac{150 + j75\ \text{kVA}}{7.39\ \text{kV}\angle{-120°}}\right| \\ &= |22.70\ \text{A}\angle 87°| \\ &= 22.70\ \text{A} \quad (23\ \text{A}) \end{aligned}$$

The answer is (C).

68. The equation for capacitive reactance per unit length is

Phasor Transforms of Sinusoids

$$X_C = \frac{1}{2\pi f C_l}$$

The capacitance per unit length is unknown in this equation. It can be found from the following single-phase capacitance equation.

$$\begin{aligned} C_l &= \frac{\pi\epsilon_0}{\ln\frac{D}{r}} \\ &= \frac{\pi\left(8.854\times 10^{-12}\ \frac{\text{F}}{\text{m}}\right)}{\ln\frac{(0.3\ \text{m})\left(1000\ \frac{\text{mm}}{\text{m}}\right)}{(20\ \text{mil})\left(0.0254\ \frac{\text{mm}}{\text{mil}}\right)}} \\ &= 4.36\times 10^{-12}\ \text{F/m} \end{aligned}$$

Substitute into the equation for capacitive reactance per unit length.

$$\begin{aligned} X_{C,l} &= \frac{1}{2\pi f C_l} \\ &= \frac{1}{2\pi(10\times 10^9\ \text{Hz})\left(4.36\times 10^{-12}\ \frac{\text{F}}{\text{m}}\right)} \\ &= 3.65\ \Omega/\text{m} \quad (4\ \Omega/\text{m}) \end{aligned}$$

The answer is (C).

69. All of the statements, with the exception of option D, are true. The acronym *SCADA* stands for supervisory control and data acquisition.

The answer is (D).

70. Triplen harmonics include those that are a factor of 3 from the fundamental, which is 60 Hz in the United States. Therefore, option A is true. Triplen harmonics increase current and thus heating, which increases the wiring temperature. Option B is true. Delta connections keep triplen harmonics within the delta loop, preventing an impact to line quantities. Option D is true.

Unlike other harmonics, triplen harmonics do not cancel at the neutral of a wye connection. Instead, they combine together to increase the neutral current. Option C is false.

The answer is (C).

71. Power disturbances are deviations from the desired power quality. The disturbance caused by harmonics is represented by option A.

The answer is (A).

72. A short circuit is tested at the rated current. The rated current on the high-voltage side results when 7% of the rated voltage is applied.

$$\begin{aligned} \text{test voltage} &= (\text{rated voltage})(\text{per-unit impedance}) \\ &= (12{,}200\ \text{V})(0.07) \\ &= 854\ \text{V} \quad (850\ \text{V}) \end{aligned}$$

The answer is (C).

73. The current in phase A is found using the power equation on the line quantities.

3-Phase Circuits

$$\begin{aligned} S &= IV \\ I &= \frac{S}{V} \\ &= \frac{3.3\times 10^3\ \text{VA}}{220\ \text{V}} \\ &= 15\ \text{A} \end{aligned}$$

The answer is (B).

74. Lithium-ion batteries have very high energy-to-weight ratios, a slow self-discharge rate, and no memory effect when left charged for a significant period of time. However, they are higher in cost than other types.

The answer is (D).

75. Conductor size is based on 100% of the noncontinuous load plus 125% of the continuous load (*National Electrical Code* (NEC) Sec. 210.19(A)(1)). Therefore, the total load is

$$\begin{aligned} P_{\text{total}} &= L_{\text{noncontinuous}} + 1.25L_{\text{continuous}} \\ &= 3\ \text{A} + (1.25)(10\ \text{A}) \\ &= 15.5\ \text{A} \end{aligned}$$

The ampacity cannot be less than the rating of the overcurrent device, which is 20 A in this case (NEC Sec. 310.15(B) Informational Note No. 2). For a circuit with termination provisions rated for less than 100 A, the ampacity of the conductor should be based on the 60°C column of NEC Table 310.15(B)(16) (NEC Sec. 110.14(C)(1)).

Using NEC Table 310.15(B)(16), an AWG 14 conductor is initially chosen. Checking NEC Sec. 240.4(D), as is called for by the asterisk in the table, indicates that a 14-gage wire cannot have overcurrent protection greater than 15 A. Therefore, an AWG 12 conductor should be selected. Note that any derating of the conductor ampacity requires the use of a small-gage (larger-diameter) conductor. Refer to the asterisk at the bottom of NEC Table 310.15(B)(16) for the tables for the conductor correction factors for ambient temperatures other than 30°C (NEC Table 310.15(B)(2)(a) and Table 310.15(B)(2)(b)).

The answer is (B).

76. A fully charged battery should have the 12 V rated voltage available. This 12 V will pass through the internal resistance of 0.5 Ω prior to being supplied to the load. Using Ohm's law, the current is

$$\begin{aligned} V &= IR \\ I &= \frac{V}{R} = \frac{12\ \text{V}}{0.5\ \Omega + 2\ \Omega} \\ &= 4.8\ \text{A} \quad (5.0\ \text{A}) \end{aligned}$$

The answer is (A).

77. Series-wound motors have high starting torques, as illustrated.

The answer is (C).

78. The programmable logic array has a total of 24 available sites. Of those sites, 10 are utilized, indicated by the "X" marking a connection in the OR array. Therefore, the array utilization is $10/24 = 0.42$ or 42%.

The answer is (C).

79. The armature speed of a generator is the synchronous speed. The synchronous speed is

Synchronous Machines: Synchronous Speed

$$\begin{aligned} n = n_s &= \frac{120f}{p} \\ &= \frac{(120)(60\ \text{Hz})}{4\ \text{poles}} \\ &= 1800\ \text{rpm} \end{aligned}$$

The answer is (C).

80. Transformer losses are either core losses (also called iron losses) or copper losses (also called winding losses). Core losses consist of eddy current losses and hysteresis losses, neither of which depends on the transformer load.

The answer is (D).

Solutions

Exam 2: Morning Session

Content in blue refers to the *NCEES Handbook*.
Content in red is additional essential information.

81. The phase and line voltages in a delta-connected load are identical. The impedance is

Phasor Transforms of Sinusoids

$$\begin{aligned} \mathbf{Z} = \frac{\mathbf{V}}{\mathbf{I}} &= \frac{120\ \text{V}\angle 30^\circ}{10\ \text{A}\angle 5^\circ} \\ &= 12\ \Omega\angle 25^\circ \\ &= 10.88 + j5.07\ \Omega \end{aligned}$$

The reactance is 5.07 Ω, which is approximately 5.1 Ω.

The answer is (A).

82. This is a transient study, so the transient values should be used. The following equation is derived from Kirchhoff's voltage law on a generator model.

$$\begin{aligned} E_g' &= V_t + jI_{\text{load}}X_d' \\ &= 1.00\ \text{pu} + j(1.00 + j0\ \text{pu})(0.30) \\ &= 1.00 + j0.30\ \text{pu} \\ &= 1.04\ \text{pu}\angle 16.7^\circ \quad (1.04\ \text{pu}) \end{aligned}$$

The answer is (B).

83. The real power is

Complex Power

$$P = IV(\text{pf})$$

The current and power factor are the unknowns. The current is

$$\begin{aligned} \mathbf{I}_{\text{eff}} &= \mathbf{I}_{\text{rms}} \\ &= \frac{\mathbf{V}}{\mathbf{Z}} \\ &= \frac{110\ \text{V}\angle 0^\circ}{6 + j3\ \Omega} \\ &= \frac{110\ \text{V}\angle 0^\circ}{6.71\ \Omega\angle 26.6^\circ} \\ &= 16.39\ \text{A}\angle -26.6^\circ \end{aligned}$$

The power factor is the cosine of the impedance angle.

Complex Power

$$\begin{aligned} \text{pf} &= \cos\phi \\ &= \cos 26.6^\circ \\ &= 0.894 \end{aligned}$$

Substitute the calculated values and the given information into the power formula.

$$\begin{aligned} P &= IV(\text{pf}) \\ &= (16.39\ \text{A})(110\ \text{V})(0.894) \\ &= 1611.8\ \text{W} \quad (1.6\ \text{kW}) \end{aligned}$$

The answer is (C).

84. A turns ratio of 200:5 can be simplified to a value of

Single-Phase Transformer Equivalent Circuits

$$\begin{aligned} a &= \frac{N_1}{N_2} \\ &= \frac{200}{5} \\ &= 40 \end{aligned}$$

The voltage from the neutral of the wye to the ground that results in the required 5 V to trip the relays is

$$\begin{aligned} V_{\text{NG}} &= \frac{aV_{\text{trip}}}{1000\ \frac{\text{V}}{\text{kV}}} \\ &= \frac{(40)(5\ \text{V})}{1000\ \frac{\text{V}}{\text{kV}}} \\ &= 0.2\ \text{kV} \end{aligned}$$

The answer is (A).

85. The power triangle, with labels and units shown, is [Complex Power Triangle (Inductive Load)]

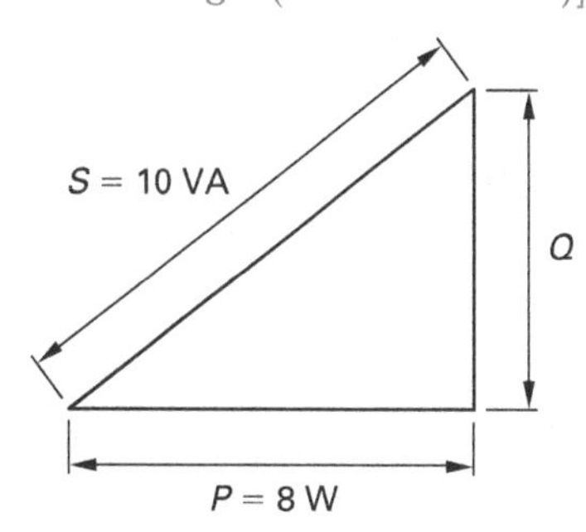

The reactive power is [Complex Power Triangle (Inductive Load)]

$$\begin{aligned} P^2 + Q^2 &= S^2 \\ Q^2 &= S^2 - P^2 \\ Q &= \sqrt{S^2 - P^2} \\ &= \sqrt{(10\ \text{VA})^2 - (8\ \text{W})^2} \\ &= 6\ \text{VAR} \end{aligned}$$

The answer is (B).

86. During the fault condition, the following current is experienced on the current transformer's secondary side.

Single-Phase Transformer Equivalent Circuits

$$a = \frac{I_p}{I_s}$$

$$\begin{aligned} I_s &= I_p\left(\frac{1}{a}\right) \\ &= (5000\ \text{A})\left(\frac{1}{100}\right) \\ &= 50\ \text{A} \end{aligned}$$

With a burden of 1.5 Ω, the secondary voltage of the current transformer is

$$\begin{aligned} V_s &= I_s R_s \\ &= (50\ \text{A})(1.5\ \Omega) \\ &= 75\ \text{V} \end{aligned}$$

The answer is (B).

87. Use the ampere-dot rule and the properties of an ideal transformer.

$$\begin{aligned} N_p\mathbf{I}_p &= N_{s1}\mathbf{I}_{s1} + N_{s2}\mathbf{I}_{s2} \\ 12\mathbf{I}_p &= (1)(5\ \text{A}\angle 30°) + (1)(6\ \text{A}\angle 10°) \\ &= (4.33 + j2.5\ \text{A}) + (5.91 + j1.04\ \text{A}) \\ &= 10.24 + j3.54\ \text{A} \\ \mathbf{I}_p &= \frac{10.24 + j3.54\ \text{A}}{12} \\ &= 0.853 + j0.295\ \text{A} \\ &= 0.90\ \text{A}\angle 19.1° \quad (0.90\ \text{A}\angle 19°) \end{aligned}$$

The answer is (B).

88. The impedance angle associated with the power factor is

Complex Power

$$\begin{aligned} \text{pf} &= \cos\theta \\ \phi &= \arccos \text{pf} \\ &= \arccos 0.8 \\ &= 36.9° \end{aligned}$$

The load impedance is

$$\begin{aligned} \mathbf{Z} &= \mathbf{R} + \mathbf{jX} \\ &= 50\ \Omega + jX\ \Omega \end{aligned}$$

The reactance is found using the known impedance angle.

$$\phi = \arctan\frac{X}{R}$$

$$\begin{aligned} \tan\phi &= \frac{X}{R} \\ X &= R\tan\phi \\ &= (50\ \Omega)\tan 36.9° \\ &= 37.5\ \Omega \quad (38\ \Omega) \end{aligned}$$

The answer is (C).

89. The distance to be calculated is shown in the figure as x.

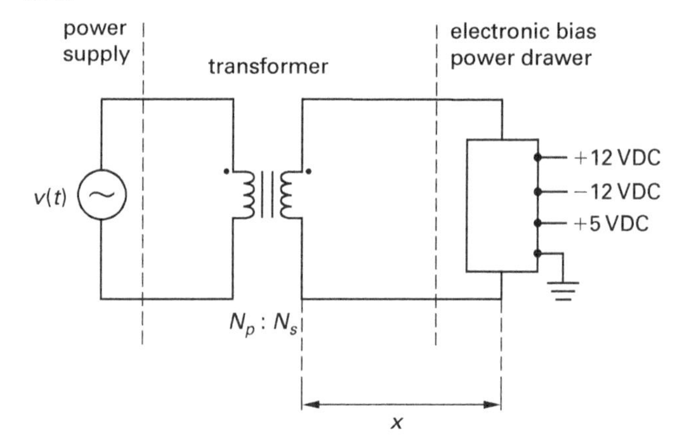

The resistance per length is given. To obtain the length, the total resistance must be determined. The maximum current is 3 A. The total resistance is

$$\begin{aligned} R_{\text{drop}} &= \frac{V_{\text{drop}}}{I} = \frac{7.6\ \text{V}}{3\ \text{A}} \\ &= 2.53\ \Omega \end{aligned}$$

The distance, x, is half the distance that the current must travel in the wire (once along each conductor). Therefore, the distance, in meters, is

$$x = \left(\frac{1}{2}\right)\left(\frac{2.53\ \Omega}{\frac{2.7\ \Omega}{1000\ \text{ft}}}\right)\left(0.3048\ \frac{\text{m}}{\text{ft}}\right)$$
$$= 142.80\ \text{m} \quad (140\ \text{m})$$

The answer is (A).

90. Using any transformer at higher frequencies makes saturation less likely. Option A is false.

Protective current transformers are designed to function at higher fault currents than instrument current transformers. The two cannot be swapped. Further, the time response is likely delayed or eliminated if an instrument CT is used in lieu of a protective CT. Option B is false.

Current transformers generate high voltage when the secondary windings are open, because the primary is sensing current flow and remains online. Therefore, the secondary must be shorted and connected to a safety ground. Option C is true.

The answer is (C).

91. The force on a current-carrying conductor within a magnetic field is given by the Lorentz force.

$$\mathbf{F} = q\mathbf{v}\times\mathbf{B}$$

The term $\mathbf{v}$ is the velocity of the charge. Using the flow of electrons, the situation is

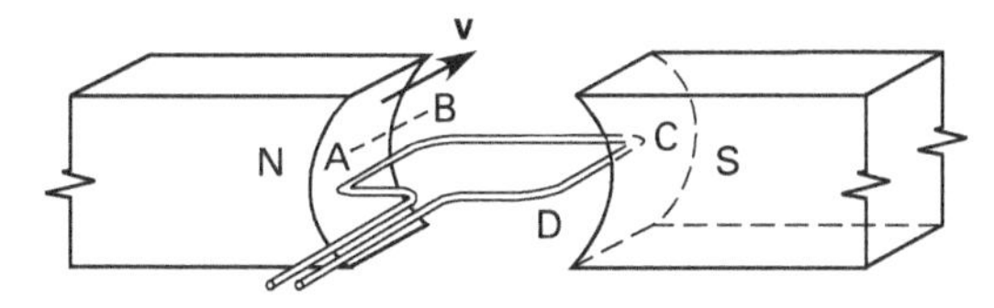

Focusing only on the wire within the magnetic field, from points A to B, the picture is

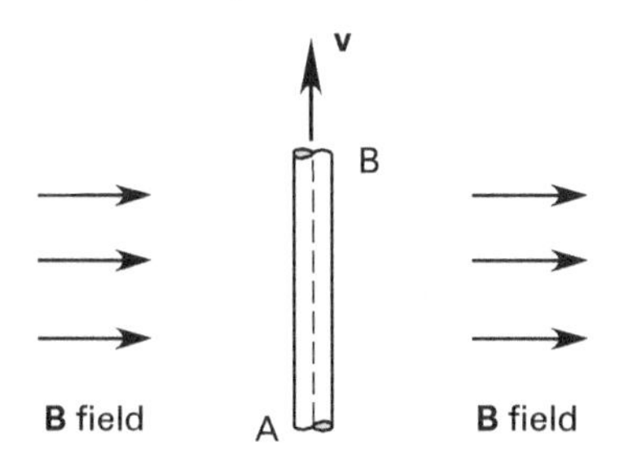

The $\mathbf{v} \times \mathbf{B}$ term gives

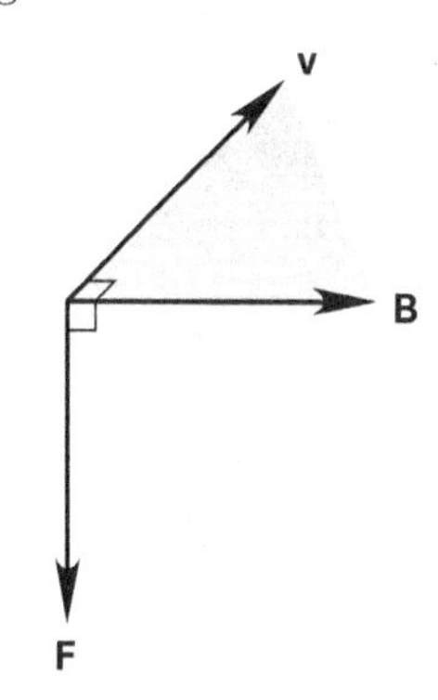

Accounting for the negative charge changes the picture to

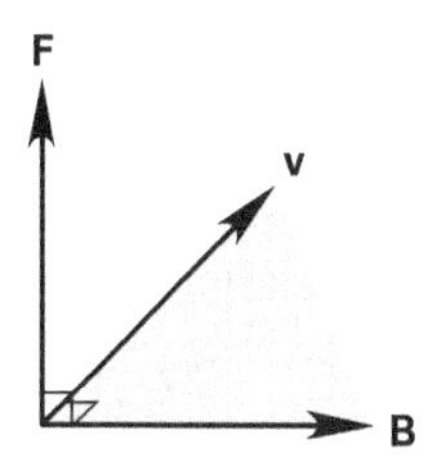

Therefore, the leftmost portion of the coil moves upward. The rightmost portion moves downward, or clockwise. The source cannot be alternating without changing the direction of this torque. The capacitor is not an energy source. The DC source is necessary to cause electron flow in the direction indicated.

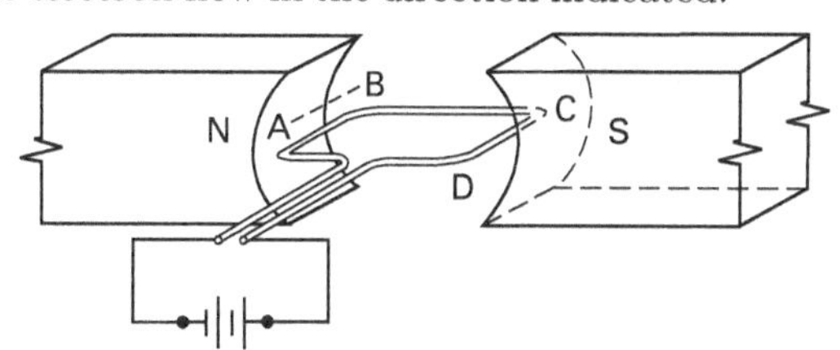

The answer is (B).

92. The current is calculated using the apparent power.

3-Phase Circuits

$$S = \sqrt{3}\,IE$$
$$I = \frac{S}{\sqrt{3}\,E}$$
$$= \frac{3\times 10^6\ \text{VA}}{\sqrt{3}\left(11\times 10^3\ \text{V}\right)}$$
$$= 157\ \text{A} \quad (160\ \text{A})$$

The answer is (B).

93. With a voltage of 0 V between points B and D, the Wheatstone bridge is balanced.

$$\frac{R_{AB}}{R_{BC}} = \frac{R_{AD}}{R_{DC}}$$

Substitute the given values and the unknown resistance.

$$\frac{R_x}{5\ \Omega} = \frac{30\ \Omega}{15\ \Omega}$$

$$R_x = (5\ \Omega)\left(\frac{30\ \Omega}{15\ \Omega}\right) = 10\ \Omega$$

The answer is (B).

94. Since the motor is not online, the generator acts as the fault's only source of power. The per-unit impedance of the generator is 0.10 on a 5 MVA base. The per-unit impedance of the transmission line must be calculated.

$$Z_{\text{pu,line}} = \frac{Z_{\text{act,line}}}{Z_{\text{base}}}$$

The base impedance is

Per Unit System

$$Z_{\text{base}} = \frac{V_{\text{base}}^2}{S_{\text{base}}} = \frac{\left((13.8\ \text{kV})\left(10^3\ \frac{\text{V}}{\text{kV}}\right)\right)^2}{(5\ \text{MVA})\left(10^6\ \frac{\text{V}}{\text{MV}}\right)} = 38.1\ \Omega$$

Substitute the base impedance into the per-unit line impedance equation.

$$Z_{\text{pu,line}} = \frac{Z_{\text{act,line}}}{Z_{\text{base}}} = \frac{2\ \Omega + j7\ \Omega}{38.1\ \Omega} = 0.05 + j0.18\ \text{pu}$$

The situation is as indicated in the following one-line diagram.

j0.10 pu — 0.05 — j0.18 pu — F1

Combine the generator and the line impedances.

$$\begin{aligned} Z_{\text{pu,total}} &= Z_{\text{pu,gen}} + Z_{\text{pu,line}} \\ &= j0.10\ \text{pu} + (0.05 + j0.18\ \text{pu}) \\ &= 0.05 + j0.28\ \text{pu} \\ &= 0.28\ \text{pu}\angle 80^\circ \quad (0.3\ \text{pu}\angle 80^\circ) \end{aligned}$$

The resistance is negligible and could have been ignored. The angle is not important to the maximum short circuit value and is shown for clarity only.

Apply the MVA method to determine the short-circuit power.

$$S_{\text{sc}} = \frac{S_{\text{base}}}{Z_{\text{pu}}} = \frac{5\ \text{MVA}}{0.3\ \text{pu}} = 16.7\ \text{MVA}$$

The short-circuit current is

$$I_{\text{sc}} = \frac{S_{\text{sc}}}{\sqrt{3}\,V_{\text{base}}} = \frac{(16.7\ \text{MVA})\left(10^6\ \frac{\text{VA}}{\text{MVA}}\right)}{\sqrt{3}(13.8\ \text{kV})\left(10^3\ \frac{\text{V}}{\text{kV}}\right)} = 698.7\ \text{A} \quad (0.7\ \text{kA})$$

The answer is (B).

95. Voltmeter A is a DC voltmeter, which reads the average of the input signal. The average is determined directly from the graph.

$$V_{\text{reading,A}} = V_{\text{ave}} = \frac{\text{area}}{\text{period}} = \frac{\left(\frac{1}{2}\right)(1\ \text{ms})(5\ \text{V})}{2\ \text{ms}} = 1.25\ \text{V}$$

For a sinusoid, the average and rms values are related as

Average Value

$$V_{\text{ave}} = \frac{2}{\pi}V_{\text{peak}} \quad \text{[for a rectified sinusoid]} \quad \text{[I]}$$

Effective or RMS Values

$$X_{\text{rms}} = \frac{X_{\text{max}}}{\sqrt{2}}$$

$$V_{\text{rms}} = \frac{1}{\sqrt{2}}V_{\text{peak}} \quad \text{[II]}$$

$$V_{\text{rms}} = 1.11\,V_{\text{ave}} \quad \text{[III]}$$

Voltmeter B uses a half-wave rectifier (i.e., a single diode). Therefore, the value of the output is one-half the value given in Eq. I.

$$V_{\text{ave}} = \frac{1}{\pi}V_{\text{peak}} \quad \text{[IV]}$$

Voltmeter B reads the rms value. Relate the value of the peak voltage obtained from Eq. II and Eq. IV.

$$V_{\text{peak}} = \sqrt{2}\,V_{\text{rms}} \quad \text{[from II]}$$
$$V_{\text{peak}} = \pi V_{\text{ave}} \quad \text{[from IV]}$$

Combine the two to relate the average value to the rms value, which is what voltmeter B is designed to read. (This has already been accomplished in Eq. III for full-wave rectification.)

$$\begin{aligned} V_{\text{rms}} &= \frac{\pi}{\sqrt{2}} V_{\text{ave}} \\ &= 2.22\,V_{\text{ave}} \quad \text{[V]} \end{aligned}$$

The average voltage was determined as 1.25 V. Substituting this value into Eq. V gives the reading on voltmeter B.

$$\begin{aligned} V_{\text{reading,B}} = V_{\text{rms}} &= 2.22\,V_{\text{ave}} \\ &= (2.22)(1.25\text{ V}) \\ &= 2.78\text{ V} \end{aligned}$$

Compare the reading of voltmeter B to voltmeter A.

$$\begin{aligned} \frac{\text{voltmeter B}}{\text{voltmeter A}} &= \frac{2.78\text{ V}}{1.25\text{ V}} \\ &= 2.22 \quad (2.0) \end{aligned}$$

The answer is also apparent from Eq. V without calculating the actual meter readings. When using meter readings directly from an instrument, errors can be introduced by failing to consider the instrument's method for determining the value of a signal, or, for a digital multimeter, the setting of the multimeter when measurements are taken.

The answer is (D).

96. From NEC Sec. 250.53(G), a rod and pipe electrode must be buried at least 8 ft deep. If that depth is not possible due to rock, the electrode can be placed at a 45° angle. If neither of those options is viable, the electrode should be buried at least 30 in deep.

Use keywords from the problem as starting points in the *NEC Handbook* index to help narrow the search for the correct article.

The answer is (D).

97. The NEC permits the marking of cables using metric areas as long as the AWG or circular mil area is also marked. Since the circular mil area determines the AWG size designation, the circular mil area must be determined. Use unit analysis to determine the area in square inches.

$$\begin{aligned} A_{\text{in}^2} &= \frac{5.00\text{ mm}^2}{\left(10\,\frac{\text{mm}}{\text{cm}}\right)^2\left(2.54\,\frac{\text{cm}}{\text{in}}\right)^2} \\ &= 0.0078\text{ in}^2 \end{aligned}$$

Find the diameter of the conductor or cable.

$$\begin{aligned} A_{\text{in}^2} &= \pi r^2 = \pi\left(\frac{d}{2}\right)^2 \\ &= \left(\frac{\pi}{4}\right)d^2 \end{aligned}$$

$$\begin{aligned} d^2 &= \frac{4A_{\text{in}^2}}{\pi} \\ d &= \sqrt{\frac{4A_{\text{in}^2}}{\pi}} \\ &= \sqrt{\frac{(4)(0.0078\text{ in}^2)}{\pi}} \\ &= 0.0993\text{ in} \end{aligned}$$

The area of a conductor in circular mils is defined as the diameter of the conductor in mils squared. Find the circular mil area for this conductor.

$$\begin{aligned} A_{\text{cmil}} &= \left(\frac{d_{\text{in}}}{0.001}\right)^2 \\ &= \left(\frac{0.0993\text{ in}}{0.001}\right)^2 \\ &= 9860\text{ cmil} \end{aligned}$$

Using the information from the given table, conductors of 10 gage or greater have the required minimum area.

The answer is (B).

98. The average value of a waveform is

Average Value

$$X_{\text{ave}} = \frac{1}{T}\int_0^T x(t)\,dt$$
$$I_{\text{ave}} = \frac{1}{T}\int_0^T i(t)\,dt$$

According to the problem statement, the average of this waveform should be 17% of the peak value of 1.0.

Substitute the known quantities into the average waveform equation.

$$\begin{aligned} I_{\text{ave}} &= \frac{1}{T}\int_0^T i(t)\,dt \\ 0.17 &= \frac{1}{\pi}\int_x^{180^\circ} \sin\phi\,dt \\ &= -\frac{1}{\pi}\cos\phi\Big|_x^{180^\circ} \\ &= -\frac{1}{\pi}(\cos 180^\circ - \cos x) \\ &= -\frac{1}{\pi}(-1 - \cos x) \\ &= \frac{1}{\pi} + \frac{\cos x}{\pi} \\ \cos x &= -0.47 \\ x &= \arccos - 0.47 \\ &= (2.06\ \text{rad})\left(\frac{180^\circ}{\pi}\right) \\ &= 118^\circ \quad (120^\circ) \end{aligned}$$

The answer is (D).

99. Low voltage is defined as 24 V or less (see NEC Sec. 551.2). High voltage is defined as 1000 V or greater (see NEC Sec. 490.2). Therefore, a medium-voltage circuit is somewhere in between, and the data in NEC Table 310.15(B)(16) for conductors rated 0–2000 V is applicable. For a temperature of 90°C, the THHN copper conductor has a maximum ampacity of 40 A for a size of AWG 10.

Conductors may have higher temperature ratings than their termination connections. However, in such cases, NEC Sec. 110.14(C)(1)(a) requires termination connections for circuits rated for 100 A or less, or requires using AWG 14 through AWG 1 wires, to limit the conductor to the ampacity ratings listed in the 60°C column. Using NEC Table 310.15(B)(16), the maximum current for an AWG 10 wire at 60°C is 30 A after derating. (Most circuits are in this amp range, so the 60°C column is most often used in calculations.)

The data represented in NEC Table 310.15(B)(16) is based on an ambient temperature of 30°C. Correction factors for other temperatures are given in NEC Sec. 310.15(B)(2)(a). For 50°C, the correction factor is 0.82 for the 90°C column. Therefore, the maximum permitted ampacity is

$$\begin{aligned} \text{ampacity} &= (\text{allowed conductor rating})(\text{correction factor}) \\ &= (40\ \text{A})(0.82) \\ &= 32.8\ \text{A} \quad (33\ \text{A}) \end{aligned}$$

This exceeds the 60°C terminal allowed rating of 30 A. Therefore, the 30 A limit must be used.

The answer is (B).

100. From NEC Sec. 240.4(D), the maximum overcurrent protection must not exceed 20 A for 12 AWG wire.

Use keywords from the problem as starting points in the *NEC Handbook* index to help narrow the search for the correct article.

The answer is (D).

101. The branch-circuit conductor ampacity, prior to the application of any adjustment or derating factors, is required by NEC Sec. 210.19(A)(1)(a) to be equal to or greater than the noncontinuous load plus 125% of the continuous load.

$$A_{\text{total}} = A_{\text{noncontinuous}} + 1.25A_{\text{continuous}}$$

Substitute the given information.

$$\begin{aligned} A_{\text{total}} &= A_{\text{noncontinuous}} + 1.25A_{\text{continuous}} \\ &= 5\ \text{A} + (1.25)(20\ \text{A}) \\ &= 30\ \text{A} \end{aligned}$$

The answer is (D).

102. For a single-phase alternating motor rated at 5 hp and 230 V, the current is 28 A [NEC Table 430.248]. Conductors supplying such loads must have 125% of the motor's full-load current rating per NEC Sec. 430.22. Therefore, letting AF represent the adjustment factor,

$$\begin{aligned} I_{\text{motor}} &= I_{\text{rated}}\text{AF} \\ &= (28\ \text{A})(1.25) \\ &= 35\ \text{A} \end{aligned}$$

The load for the resistance heater is

$$\begin{aligned} I_{\text{heater}} &= \frac{P}{V} \\ &= \frac{500\ \text{W}}{240\ \text{V}} \\ &= 2.08\ \text{A} \end{aligned}$$

The total load is

$$\begin{aligned} I_{\text{total}} &= I_{\text{motor}} + I_{\text{heater}} \\ &= 35\ \text{A} + 2.08\ \text{A} \\ &= 37.08\ \text{A} \quad (38\ \text{A}) \end{aligned}$$

Since the total load is 37.08 A, the minimum ampacity is 38 A. According to NEC Table 310.15(B)(16), no temperature correction factor is required for an ambient temperature of 30°C. For a THHN wire and a minimum ampacity of 38 A, the required wire size is 10 AWG.

The answer is (C).

103. Per NEC Sec. 210.20(A), the overcurrent protection for branch circuit conductors and equipment is set at not less than the noncontinuous load plus 125% of the continuous load.

$$A_{\text{OPCD}} = A_{\text{noncontinuous}} + 1.25A_{\text{continuous}}$$

Substituting the given information, the overcurrent protection is

$$\begin{aligned} A_{\text{OPCD}} &= A_{\text{noncontinuous}} + 1.25A_{\text{continuous}} \\ &= 8\ \text{A} + (1.25)(20\ \text{A}) \\ &= 33\ \text{A} \end{aligned}$$

Standard ampere ratings of overcurrent devices are given in NEC Table 240.6(A) as 25 A, 30 A, 35 A, 40 A, 45 A, and so on. Given that the circuit does not contain receptacles, use of the next higher standard overcurrent device rating is permitted (see NEC Sec. 240.4(B)(1)). Therefore, an OCPD at 35 A is permitted. (It is important to note that had this circuit contained two or more receptacles, the OCPD would have been set at the rating of the circuit, 30 A. See NEC Sec. 210.3 and Table 210.24 for further information.)

The answer is (C).

104. From NEC Table 430.251(B), the locked-rotor current is 363 A. The full-load current is 65 A [NEC Table 430.250]. The ratio of locked-rotor to full-load current is

$$\begin{aligned} R &= \frac{I_{\text{locked-rotor}}}{I_{\text{full-load}}} \\ &= \frac{363\ \text{A}}{65\ \text{A}} \\ &= 5.58 \quad (6.0) \end{aligned}$$

The locked-rotor current is often considered to be the starting current. Therefore, the ratio represents a factor that can be multiplied by the full-load current to find the starting current. Starting current is generally five to seven times larger than the running (full-load) current.

The answer is (A).

105. Using NEC Table 310.15(B)(16), in the row for AWG 10, the TW conductor is in the 60°C column with an allowable ampacity of 30 A. The THW conductor is in the 75°C column with an allowable ampacity of 35 A. The THHW wire appears in two columns, but the rating is given in the problem statement as 90°C. Using the 90°C rating column, the allowable ampacity for the THHW (90°C) wire is 40 A.

The table is for an ambient temperature of 30°C, so no temperature correction is required. However, the table is to be used for not more than three current-carrying conductors in a raceway. NEC Sec. 310.15(B)(3)(a) and Table 310.15(B)(3)(a) require derating factors to be used when more than three conductors are present. The NEC table is adapted as follows.

number of current-carrying conductors	percentage of values in NEC Table 310.15(B)(16) as adjusted for ambient temperature
4–6	80%
7–9	70%
10–20	50%
21–30	45%
31–40	40%
41+	35%

Since four conductors are within the cable, a derating factor of 80% is applicable. The TW cable is allowed to carry

$$\begin{aligned} A_{\text{max}} &= (\text{allowable ampacity})(\text{derating factor}) \\ &= (30\ \text{A})(0.80) \\ &= 24\ \text{A} \end{aligned}$$

By similar calculations, the THW conductor amperage is 28 A. The THHW (90°C) conductor amperage is 32 A. Therefore, only the THHW (90°C) cable meets the required conditions.

The answer is (C).

106. From the Informational Note in NEC Sec. 310.15(B), the considerations given for the values in the table include option B, option C, and option D. The NEC determines the most important and applicable factors. It is possible that values from the table may be too high for certain applications. NEC Annex B provides some additional examples of such factors.

The answer is (A).

107. The characteristic impedance is

$$\begin{aligned} Z_0 &= \sqrt{\frac{L}{C}} \\ &= \sqrt{\frac{0.37\times 10^{-6}\ \frac{\text{H}}{\text{m}}}{100\times 10^{-12}\ \frac{\text{F}}{\text{m}}}} \\ &= 60.8\ \Omega \quad (61\ \Omega) \end{aligned}$$

The answer is (C).

108. Since the source is ungrounded, the phase-to-phase voltage is the line voltage. This line voltage is also the phase-to-phase voltage on the delta and remains unchanged due to a single ground at the corner of C. Therefore, the voltage between A and C is 12.5 kV (12 kV).

The answer is (D).

109. Losses (and gains) are often tabulated in units of decibels for standard conditions because they can be added algebraically to determine the total. Therefore, a 1 km line will have 10 times the loss of a 100 m line.

$$\begin{aligned} L &= \left(\frac{0.95\ \text{dB}}{100\ \text{m}}\right)(1\ \text{km})\left(1000\ \frac{\text{m}}{\text{km}}\right) \\ &= 9.50\ \text{dB} \end{aligned}$$

The answer is (D).

110. It is mathematically correct to perform calculations as if a neutral exists, and then convert back to line quantities. Recall that the line quantities are larger in magnitude by $\sqrt{3}$ and lead the phase quantities by 30°. Therefore, use Kirchhoff's voltage law around this imaginary loop, and adjust the load voltage as if it were connected to a neutral.

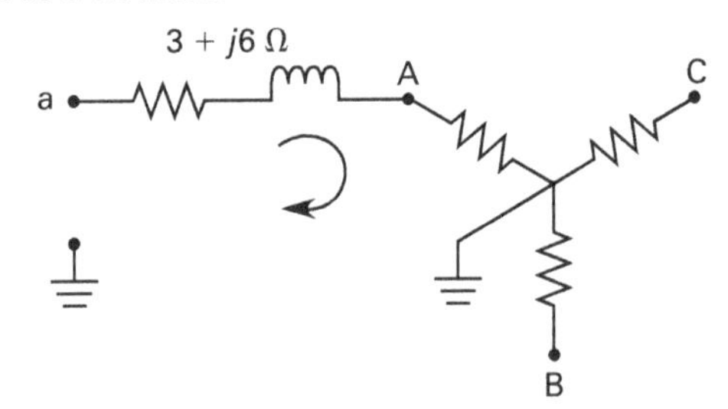

$$\begin{aligned} \mathbf{V}_{\text{an}} &= \mathbf{V}_{\text{line}} + \mathbf{V}_{\text{load}} \\ &= \mathbf{I}_{\text{a}}\mathbf{Z}_{\text{line}} + \mathbf{V}_{\text{load}} \\ &= (50\ \text{A}\angle{-30°})(3 + j6\ \Omega) + \frac{13.2\ \text{kV}\angle{-30°}}{\sqrt{3}} \\ &= 335.4\ \text{V}\angle 33.4° + 7.62 \times 10^3\ \text{V}\angle{-30°} \\ &= 7777\ \text{V}\angle 27.79° \end{aligned}$$

Convert this value to the line quantity.

$$\begin{aligned} |\mathbf{V}_{\text{ab}}| &= \sqrt{3}\,(|\mathbf{V}_{\text{an}}|) \\ &= \sqrt{3}\,(7777\ \text{V}) \\ &= 13{,}479\ \text{V} \quad (13\ \text{kV}) \end{aligned}$$

The answer is (C).

111. The normalized load impedance is determined by dividing the load impedance by the characteristic impedance.

$$\begin{aligned} z_L &= \frac{Z_L}{Z_0} \\ &= \frac{120 + j150\ \Omega}{30\ \Omega} \\ &= 4.00 + j5.00 \end{aligned}$$

The answer is (D).

112. The speed of DC motors can be changed using a variety of methods, nearly all of which are less complicated and inexpensive than the methods used for speed control in AC machines. Reversal is possible with DC machines without power switching. DC motors also continuously operate over a speed range of 8:1.

AC motors have a tendency to stall, while DC motors, power supply permitting, can provide up to five times the rated torque. Therefore, option C is false.

The answer is (C).

113. The synchronous speed (in rpm), n, is related to the output frequency, f, and the number of poles, p. The synchronous speed is

Synchronous Machines: Synchronous Speed

$$\begin{aligned} n_s &= \frac{120f}{p} \\ &= \frac{(120)(60\ \text{Hz})}{2} \\ &= 3600\ \text{rpm} \end{aligned}$$

The synchronous speed is equal to the rotational speed of the rotor in the generator.

The answer is (C).

114. The current determines the magnetic energy in the inductor, not the inductance level.

The inductance is

Reactors

$$\begin{aligned} L &= \frac{\mu N^2 A}{l} = \frac{\mu_0 \mu_r N^2 A}{l} \\ &= \frac{\left(1.2566 \times 10^{-6}\ \frac{\text{H}}{\text{m}}\right)(2000)(10)^2(3.14\ \text{m}^2)}{(6\ \text{in})\left(2.54\ \frac{\text{cm}}{\text{in}}\right)\left(\frac{1\ \text{m}}{100\ \text{cm}}\right)} \\ &= 5.18\ \text{H} \quad (5.2\ \text{H}) \end{aligned}$$

The answer is (A).

115. The desired relationship is

$$P = \frac{f_{\text{nl}} - f_{\text{sys}}}{f_{\text{droop}}}$$

A speed droop of 1.5% indicates a change in frequency of (0.015)(60 Hz) = 0.90 Hz from 0 MW to 1500 MW, that is, from no load to full load. Therefore, the slope of the speed characteristic line, or curve, is

$$f_{\text{droop}} = \frac{0.90 \text{ Hz}}{1500 \text{ MW}}$$

Rearrange and substitute this and the given information into the equation.

$$\begin{aligned} f_{\text{nl}} &= Pf_{\text{droop}} + f_{\text{sys}} \\ &= (1000 \text{ MW})\left(\frac{0.90 \text{ Hz}}{1500 \text{ MW}}\right) + 60 \text{ Hz} \\ &= 60.6 \text{ Hz} \end{aligned}$$

The answer is (B).

116. Diamagnetism is produced by electron spins in antiparallel pairs within closed shells. Ferromagnetism occurs when atomic moments arranged in domains with equal alignment are exchanged. When the exchange of atomic moments with antiparallel arrangement of equal spins occurs, the result is antiferromagnetism. Orbital or spin moments produce paramagnetism.

The answer is (B).

117. The only voltage given is the motor line voltage, V_{ml}, of 4000 V. Transforming this voltage from line to phase value,

3-Phase Circuits

$$\begin{aligned} V_{T_2\phi} &= \frac{V_{\text{ml}}}{\sqrt{3}} \\ &= \frac{4000 \text{ V}}{\sqrt{3}} \\ &= 2309.4 \text{ V} \end{aligned}$$

Transforming the phase voltage through transformer T_2 to the delta-connected side gives the transmission line voltage, V_{tl}.

$$\begin{aligned} V_{\text{tl}} &= V_{T_2\phi}\left(\frac{N_{\text{tl}}}{N_{\text{ml}}}\right) \\ &= (2309.4 \text{ V})\left(\frac{2}{1}\right) \\ &= 4618.8 \text{ V} \end{aligned}$$

The transmission line is delta connected at both ends. Therefore, the line and phase voltages are identical, meaning $V_{\text{tl}} = V_{\text{tl}\phi}$. Transforming the transmission line phase voltage through transformer T_1 to the wye-connected side gives the phase voltage on the primary side of T_1.

$$\begin{aligned} V_{T_1\phi} &= V_{\text{tl}\phi}\left(\frac{N_{\text{wye}}}{N_{\text{tl}}}\right) \\ &= (4618.8 \text{ V})\left(\frac{1}{4}\right) \\ &= 1154.7 \text{ V} \end{aligned}$$

The primary side of T_1 is wye-connected. The line voltage is

$$\begin{aligned} V_{\text{line}} &= V_{T_1\phi}\sqrt{3} \\ &= (1154.7 \text{ V})\sqrt{3} \\ &= 2000 \text{ V} \end{aligned}$$

The line voltage on the primary side of T_1 is the line voltage of the generators, which was picked as the base voltage. The base voltage is 2000 V or 2 kV.

The answer is (B).

118. From NEC Chap. 9, Table 9, the impedance for the 300 kcmil feeder is in ohms to neutral per kilometer. The total impedance for the 500 m feeder is

$$\begin{aligned} \mathbf{Z}_{\text{feeder}} &= \left(\frac{l_{\text{feeder}}}{Z_{1000 \text{ m}}}\right)\mathbf{Z}_{300 \text{ kcmil}} \\ &= \left(\frac{500 \text{ m}}{1000 \text{ m}}\right)(0.161 + j0.135 \ \Omega) \\ &= 0.0805 + j0.0675 \ \Omega \\ &= 0.105 \ \Omega\angle 40^\circ \end{aligned}$$

The voltage at PP1 is the source voltage minus the voltage drop. For a single phase, the voltage is

$$\begin{aligned} \mathbf{V}_{\text{PP1}} &= \mathbf{V}_{\text{main}} - \mathbf{I}_{\text{load}}\mathbf{Z}_{\text{feeder}} \\ &= \left(\frac{480 \text{ V}}{\sqrt{3}}\angle 0^\circ\right) - (280 \text{ A}\angle -\arccos 0.8) \\ &\quad \times (0.105 \ \Omega\angle 40^\circ) \\ &= 277.13 \text{ V}\angle 0^\circ - (280 \text{ A}\angle -36.87^\circ)(0.105 \ \Omega\angle 40^\circ) \\ &= 277.13 \text{ V}\angle 0^\circ - 29.40 \text{ V}\angle 3.13^\circ \\ &= 247.78 \text{ V}\angle -0.37^\circ \end{aligned}$$

The line (phase-to-phase) voltage is

3-Phase Circuits

$$\mathbf{V}_{\text{line}} = \sqrt{3}\,\mathbf{V}_{\text{phase}}$$

$$\begin{aligned} &= \sqrt{3}\,(247.78 \text{ V}\angle{-0.37°}) \\ &= 429.17 \text{ V}\angle{-0.37°} \end{aligned}$$

$$|\mathbf{V}_{\text{line}}| = 429 \text{ V} \quad (430 \text{ V})$$

This represents an excessive voltage drop. A larger diameter cable wire would be needed to lower the total voltage drop to the recommended 3% (NEC Sec. 215.2 (A)(1)(b) Informational Note No. 2).

The answer is (C).

119. The phase impedance is

Polar Coordinate System

$$r = |x + jy| = \sqrt{x^2 + y^2}$$

$$\theta = \arctan \frac{y}{x}$$

Algebra of Complex Numbers

$$\mathbf{Z} = r\angle\theta$$

$$\begin{aligned} \mathbf{Z}_p &= \sqrt{R^2 + X_L^2}\angle\arctan \frac{X}{R} \\ &= \sqrt{(3\ \Omega)^2 + (4\ \Omega)^2}\angle\arctan \frac{4\ \Omega}{3\ \Omega} \\ &= 5\ \Omega\angle 53.13° \quad (5\ \Omega\angle 53°) \end{aligned}$$

The answer is (A).

120. An overexcited synchronous motor can be considered a reactive power generator. This is represented by option D, where the generated voltage is greater than the terminal voltage.

The answer is (D).

Solutions

Exam 2: Afternoon Session

Content in blue refers to the *NCEES Handbook*.
Content in red is additional essential information.

121. Option D is the equivalent circuit for a shunt-wired DC motor. Option C is the equivalent circuit for a series-wired DC motor. Option B is the equivalent circuit for an AC synchronous motor. Option A is the equivalent circuit for an induction motor.

The answer is (A).

122. A Wheatstone bridge is balanced when there is no current flowing through the ammeter. The relationship between resistors in a balanced Wheatstone bridge is

$$\frac{R_{\text{var}}}{R_3} = \frac{R_2}{R_4}$$

The required value of the variable resistor is

$$\begin{aligned} R_{\text{var}} &= R_3\left(\frac{R_2}{R_4}\right) = (15\ \Omega)\left(\frac{10\ \Omega}{50\ \Omega}\right) \\ &= 3.0\ \Omega \end{aligned}$$

The answer is (A).

123. The slip percentage is calculated from the difference in no-load and full-load speed.

Percent Slip in Induction Machines

$$\begin{aligned} s &= \frac{n_s - n}{n_s} = \frac{1800\ \frac{\text{rev}}{\text{min}} - 1740\ \frac{\text{rev}}{\text{min}}}{1800\ \frac{\text{rev}}{\text{min}}} \times 100\% \\ &= 3.3\% \end{aligned}$$

The answer is (A).

124. The thermal agitation noise, also called Johnson noise, on the resistor is

$$\begin{aligned} V_{\text{noise}} &= \sqrt{4\kappa TR\Delta f} \\ &= \sqrt{\begin{array}{l}(4)\left(1.3807 \times 10^{-23}\ \frac{\text{J}}{\text{K}}\right)(25^\circ\text{C} + 273^\circ) \\ \times\left(20 \times 10^3\ \Omega\right)(1000\ \text{Hz} - 0\ \text{Hz})\end{array}} \\ &= 5.74 \times 10^{-7}\ \text{V} \quad \left(5.7 \times 10^{-7}\ \text{V}\right) \end{aligned}$$

The answer is (D).

125. The power transfer that takes place in an induction motor is illustrated as follows.

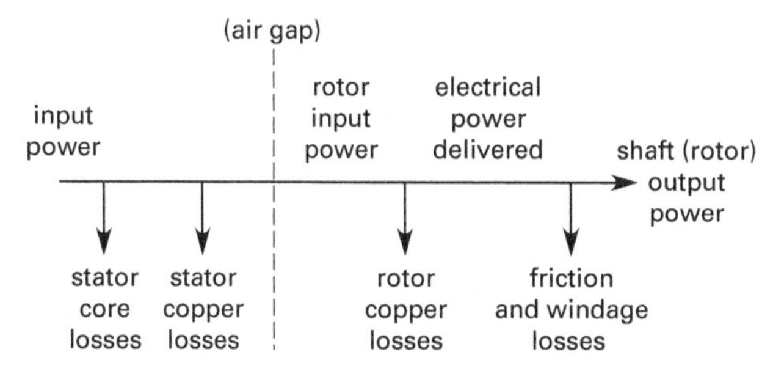

The total losses are

$$\begin{aligned} P_{\text{loss}} &= P_{\text{stator}} + P_{\text{rotor}} \\ &= 650\ \text{W} + 500\ \text{W} + 150\ \text{W} + 950\ \text{W} \\ &= 2250\ \text{W} \end{aligned}$$

The output power is 15 hp. Convert to watts. [Conversion Factor]

$$\begin{aligned} P_{\text{out}} &= (15\ \text{hp})\left(745.7\ \frac{\text{W}}{\text{hp}}\right) \\ &= 11\,185.5\ \text{W} \end{aligned}$$

Determine the input power using the calculated data.

$$\begin{aligned} P_{\text{in}} &= P_{\text{out}} + P_{\text{loss}} \\ &= 11\,185.5\ \text{W} + 2250\ \text{W} \\ &= 13\,435.5\ \text{W} \end{aligned}$$

The efficiency is

Transformer's Efficiency

$$\begin{aligned} \eta &= \frac{P_{\text{out}}}{P_{\text{in}}} \\ &= \frac{11\,185.5\ \text{W}}{13\,435.5\ \text{W}} \\ &= 0.833 \quad (80\%) \end{aligned}$$

The answer is (C).

126. VFD controllers often use insulated gate bipolar junction transistors. Option A is true. VFDs apply low frequency during the start-up cycle to minimize the starting current. Option B is true. VFD controllers use modified sine signals (often pulse width modulation,

PWM) that result in harmonics. Option C is true. The volts/hertz ratio is generally constant in order to avoid exceeding the rated torque. Therefore, option D is false.

The answer is (D).

127. The coercive force, H_c, is the magnitude of the magnetizing force where B is zero and the residual flux is removed from the material, that is, point 3.

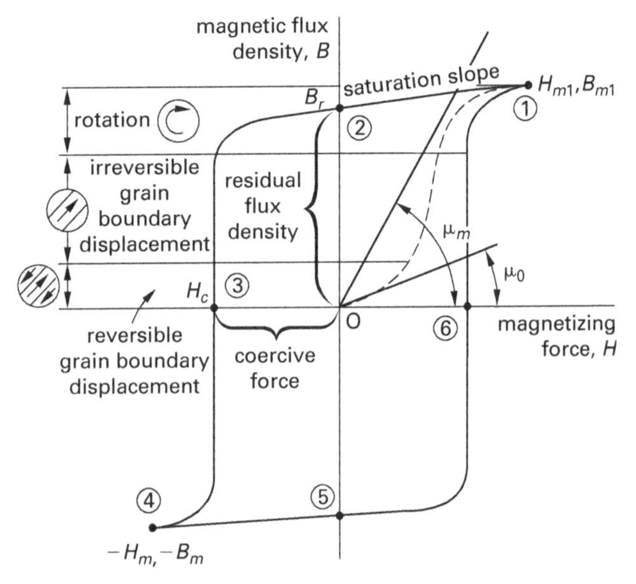

The answer is (C).

128. The magnetomotive force is

$$F_{\text{mmf}} = IN = \phi\mathfrak{R}$$

$$I = \frac{\phi\mathfrak{R}}{N}$$

The flux and number of turns is known. The reluctance is a combination of the cast-iron toroid permeability and the air-gap permeability.

$$\begin{aligned}
\mathfrak{R} &= \sum \frac{l}{\mu A} = \frac{1}{\mu_0 A}\sum \frac{l}{\mu_r} \\
&= \left(\frac{1}{\left(4\pi \times 10^{-7}\ \frac{\text{Wb}}{\text{A}\cdot\text{m}}\right)(0.0045\ \text{m}^2)}\right)\left(\frac{l_{\text{toroid}}}{\mu_{\text{iron}}} + \frac{l_{\text{air-gap}}}{\mu_{\text{air}}}\right) \\
&= \left(1.77 \times 10^8\ \frac{\text{A}}{\text{Wb}\cdot\text{m}}\right)\left(\frac{l_{\text{toroid}}}{2000} + \frac{0.001\ \text{m}}{1}\right)
\end{aligned}$$

The length of the toroid path is

$$\begin{aligned}
l_{\text{toroid}} &= \pi d \\
&= \pi(0.076\ \text{m}) \\
&= 0.24\ \text{m}
\end{aligned}$$

Substitute the length of the toroid into the reluctance equation.

$$\begin{aligned}
\mathfrak{R} &= \left(1.77 \times 10^8\ \frac{\text{A}}{\text{Wb}\cdot\text{m}}\right)\left(\frac{l_{\text{toroid}}}{2000} + \frac{0.001\ \text{m}}{1}\right) \\
&= \left(1.77 \times 10^8\ \frac{\text{A}}{\text{Wb}\cdot\text{m}}\right)\left(\frac{0.24\ \text{m}}{2000} + \frac{0.001\ \text{m}}{1}\right) \\
&= 1.98 \times 10^5\ \text{A/Wb}
\end{aligned}$$

Solve for the current.

$$\begin{aligned}
I &= \frac{\phi\mathfrak{R}}{N} \\
&= \frac{(0.01\ \text{Wb})\left(1.98 \times 10^5\ \frac{\text{A}}{\text{Wb}}\right)}{200} \\
&= 9.9\ \text{A} \quad (10\ \text{A})
\end{aligned}$$

The answer is (C).

129. A GFCI senses the magnetic flux combination of the incoming power conductor and the neutral conductor. When a differential current exists due to a ground, the magnetic fluxes of the power conductor and of the neutral conductor will not sum to approximately zero, and a net current on the order of 4 mA to 6 mA flows in the device.

If a phase-to-neutral short circuit occurs, the magnetic flux imbalance does not occur, and the GFCI is unable to sense the problem.

The answer is (D).

130. NEC Sec. 430.6 specifies the use of tables to determine the current to be used in follow-on determinations. There are exceptions: (1) multispeed motors, which undergo a change in current with the change in speed, (2) shaded-pole and capacitor-type motors, which undergo a change in current caused by reactive components, and (3) devices marked with both horsepower and ampere ratings. In all three cases, the ampere rating marked on the nameplate should be used.

Therefore, for the motor given in the problem, the full-load current value of 6.0 A specified on the nameplate should be used.

The answer is (C).

131. The skin effect, which is the tendency of alternating currents to flow near the surface of the conductor, is more pronounced the higher the frequency. The skin effect increases the AC resistance and decreases the internal inductance.

The answer is (C).

132. The fault current is 2000 A, and the minimum pickup is 333 A. The multiple of the pickup current is

$$\begin{aligned} x_{\text{multiple}} &= \frac{I_{\text{fault}}}{I_{\text{pickup}}} \\ &= \frac{2000 \text{ A}}{333 \text{ A}} \\ &= 6 \end{aligned}$$

To achieve a trip time of 0.5 s, the time dial must be set at 2.

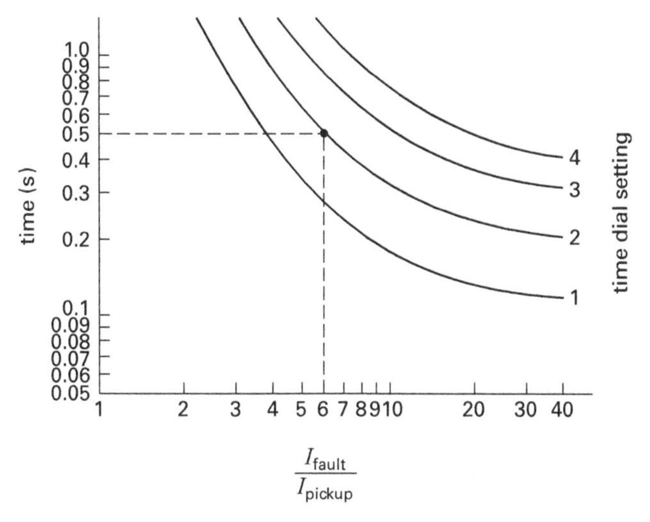

The answer is (B).

133. The characteristic impedance is

Transmission Line Models

$$\begin{aligned} Z_0 &= \sqrt{\frac{Z_l}{Y_l}} \\ &= \sqrt{\frac{0.85 \ \frac{\Omega}{\text{m}}}{7.00 \times 10^{-6} \ \frac{\text{S}}{\text{m}}}} \\ &= 348 \ \Omega \quad (350 \ \Omega) \end{aligned}$$

The answer is (C).

134. Find the magnitude of the fault current for a 0 Ω fault resistance.

$$\begin{aligned} V &= I_{\text{fault}} Z \\ &= I_{\text{fault}} |jX_s| \\ X_s &= \frac{V}{I_{\text{fault}}} \\ &= \frac{120 \text{ V}}{10{,}000 \text{ A}} \\ &= 0.012 \ \Omega \end{aligned}$$

Change the fault resistance to 0.5 Ω to find the new fault current.

$$\begin{aligned} V &= I_{\text{fault}} Z \\ &= I_{\text{fault}} |R_{\text{fault}} + jX_s| \\ I_{\text{fault}} &= \frac{V}{|R_{\text{fault}} + jX_s|} \\ &= \frac{120 \text{ V}}{|0.5 + j0.012 \ \Omega|} \\ &= \frac{120 \text{ V}}{|0.5\angle 1.37^\circ \ \Omega|} \\ &= \frac{120 \text{ V}}{0.5 \ \Omega} \\ &= 240 \text{ A} \end{aligned}$$

The answer is (A).

135. The reflection coefficient, Γ, in terms of the standing-wave ratio (SWR), is

$$\begin{aligned} \Gamma &= \frac{\text{SWR} - 1}{\text{SWR} + 1} \\ &= \frac{2 - 1}{2 + 1} \\ &= 0.33 \end{aligned}$$

The fraction of incident power reflected is Γ^2. Therefore,

$$\begin{aligned} \Gamma^2 &= (0.33)^2 \\ &= 0.11 \quad (0.1) \end{aligned}$$

The answer is (A).

136. Since the transformer is ideal, the phase angle does not change when reflected. The effective primary impedance is

Single-Phase Transformer Equivalent Circuits

$$\begin{aligned} \mathbf{Z}_{\text{effective}} &= \mathbf{Z}_{\text{reflected}} = \mathbf{Z}_p + a^2\mathbf{Z}_s \\ &= 5 \ \Omega\angle 0^\circ + (3)^2(7 + j7 \ \Omega) \\ &= 5 \ \Omega\angle 0^\circ + (9)(9.9 \ \Omega\angle 45^\circ) \\ &= 5 \ \Omega\angle 0^\circ + 89.1 \ \Omega\angle 45^\circ \\ &= 92.7 \ \Omega\angle 42.8^\circ \quad (90 \ \Omega\angle 43^\circ) \end{aligned}$$

The answer is (C).

137. Knowing the value of the line impedance, the fault current can be found.

$$\begin{aligned} V &= I_{\text{fault}} Z \\ I_{\text{fault}} &= \frac{V}{Z} \end{aligned}$$

The type of connection was not specified. Nevertheless, $\sqrt{3}$ is required in the following formula for either a wye or delta connection to determine the three-phase fault current.

$$|I_{\text{fault}}| = \left| \frac{\frac{V_l}{\sqrt{3}}}{Z} \right|$$
$$= \left| \frac{\frac{13.2 \times 10^3 \text{ V}}{\sqrt{3}}}{10\ \Omega\angle 85^\circ + 15\ \Omega\angle 75^\circ + 12\ \Omega\angle 70^\circ} \right|$$
$$= \left| \frac{7621.02 \text{ V}}{36.8\ \Omega\angle 76.1^\circ} \right|$$
$$= 207.09 \text{ A} \quad (210 \text{ A})$$

The answer is (A).

138. According to NEC Sec. 430.32(A)(1), a separate overload device should be rated at 115% of the motor's full-load current rating. The full-load current rating for a 20 hp, 208 V motor is 59.4 A [NEC Table 430.250]. The current is

$$I_{\text{OL}} = (1.15)(59.4 \text{ A})$$
$$= 68.3 \text{ A}$$

Standard fuse ratings are given in NEC Table 240.6(A) as 60 A, 70 A, and so on. To avoid exceeding requirements, the 60 A fuse must be selected.

The answer is (B).

139. From IEEE Standard C37.2, device #27 is an undervoltage relay. The "–2" indicates the second relay in the system with this number. The "*a*" indicates a contact that is normally open when the main device is in the standard reference condition. The standard reference condition for a relay is "de-energized."

The answer is (B).

140. Generally, the resistance, R, is negligible compared to the reactance, X. Therefore, $Z \approx X$. In the per-unit system, the conversion from one impedance base to another when the voltages are the same is given by the following equation.

Per Unit System

$$Z_{\text{pu,new}} = Z_{\text{pu,old}} \left(\frac{S_{\text{base,new}}}{S_{\text{base,old}}} \right)$$

Therefore, for generator 1,

$$X\% = (10\%) \left(\frac{20 \text{ MVA}}{20 \text{ MVA}} \right) = 10\%$$

For generator 2,

$$X\% = (8\%) \left(\frac{20 \text{ MVA}}{10 \text{ MVA}} \right) = 16\%$$

For the transformer,

$$X\% = (6\%) \left(\frac{20 \text{ MVA}}{30 \text{ MVA}} \right) = 4\%$$

The base voltage in this region of the system is 66 kV. For the transmission line,

$$Z_{\text{pu}} = \frac{Z_{\text{actual}}}{Z_{\text{base}}}$$
$$= \frac{Z_{\text{actual}}}{\frac{V_{\text{base}}^2}{S_{\text{base}}}}$$
$$= Z_{\text{actual}} \left(\frac{S_{\text{base}}}{V_{\text{base}}^2} \right)$$
$$= (8 + j80\ \Omega) \left(\frac{20 \times 10^6 \text{ VA}}{(66 \times 10^3 \text{ V})^2} \right)$$
$$= 0.037 + j0.37 \text{ pu}$$

Convert to percent impedance.

$$Z\% = (0.037 + j0.37_{\text{pu}}) \times 100\%$$
$$= 3.7 + j37\%$$

The result is

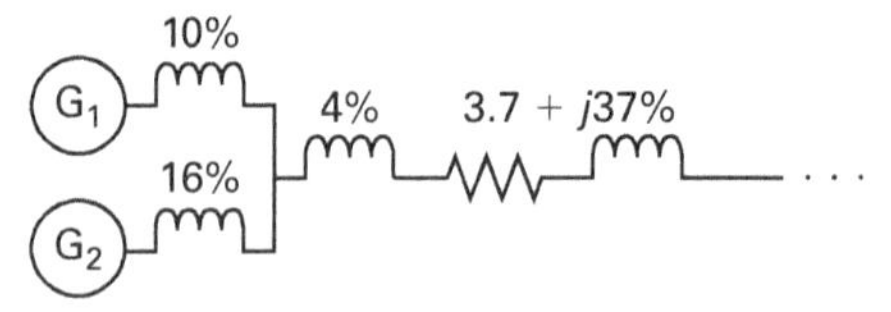

The answer is (D).

141. The short-circuit power, S_{sc} (also called the fault power), is

$$S_{\text{sc}} = \frac{S_{\text{base}}}{Z_{\text{pu}}}$$

Using a 50 kVA base (any base may be used), the per-unit impedance for each distribution element is determined from

Per Unit System

$$Z_{\text{pu,new}} = Z_{\text{pu,old}}\left(\frac{S_{\text{base,new}}}{S_{\text{base,old}}}\right)$$

Noting that $Z_{pu} \approx X_{pu}$ and using the conversion from one base to another,

$$\begin{aligned} X_{\text{generator 1}} &= (10\%)\left(\frac{50 \text{ kVA}}{30 \text{ kVA}}\right) \\ &= 16.7\% \quad (0.17 \text{ pu}) \\ X_{\text{generator 2}} &= (8\%)\left(\frac{50 \text{ kVA}}{20 \text{ kVA}}\right) \\ &= 20\% \quad (0.20 \text{ pu}) \\ X_{\text{transformer}} &= (5\%)\left(\frac{50 \text{ kVA}}{50 \text{ kVA}}\right) \\ &= 5\% \quad (0.05 \text{ pu}) \end{aligned}$$

For a fault at point A, the transmission line impedance does not impact the result, nor is it significant in either case. Therefore, the system can now be represented as follows.

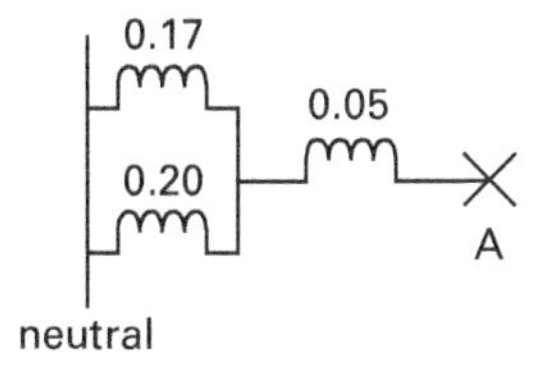

The per-unit impedance is

$$\begin{aligned} Z_{\text{pu,system}} &= X_{\text{pu,system}} \\ &= \frac{X_{\text{generator 1}}X_{\text{generator 2}}}{X_{\text{generator 1}} + X_{\text{generator 2}}} + X_{\text{transformer}} \\ &= \frac{(0.17)(0.20)}{0.17 + 0.20} + 0.05 \\ &= 0.14 \end{aligned}$$

The short-circuit power, or fault power, is

$$\begin{aligned} S_{\text{sc}} &= \frac{S_{\text{base}}}{Z_{\text{pu}}} \\ &= \frac{50 \times 10^3 \text{ VA}}{(0.14)\left(1000 \ \frac{\text{VA}}{\text{kVA}}\right)} \\ &= 357.1 \text{ kVA} \quad (400 \text{ kVA}) \end{aligned}$$

The answer is (C).

142. Since the question involves the contribution of the generator, use the generator ratings as the base values. The apparent power base is then 30 MVA. The voltage base is the voltage of the system whose desired current is under consideration. In this problem, the voltage base is 11 kV. (If one desired the current on the secondary side of the transformer, the base voltage would be that of the secondary of the transformer.)

Convert the transformer impedance using the base values.

Per Unit System

$$\begin{aligned} Z_{\text{pu,new}} &= Z_{\text{pu,old}}\left(\frac{V_{\text{base,old}}}{V_{\text{base,new}}}\right)^2\left(\frac{S_{\text{base,new}}}{S_{\text{base,old}}}\right) \\ &= (0.07 \text{ pu})\left(\frac{11 \text{ kV}}{11 \text{ kV}}\right)^2\left(\frac{30 \text{ MVA}}{40 \text{ MVA}}\right) \\ &= 0.052 \text{ pu} \end{aligned}$$

The short-circuit power is

$$\begin{aligned} S_{\text{sc}} &= \frac{S_{\text{base}}}{Z_{\text{pu}}} \\ &= \frac{S_{\text{base}}}{Z_{\text{gen}} + Z_{\text{trans}}} \\ &= \frac{30 \text{ MVA}}{0.20 \text{ pu} + 0.052 \text{ pu}} \\ &= 119.05 \text{ MVA} \quad (120 \text{ MVA}) \end{aligned}$$

The answer is (B).

143. According to the Wien displacement law, the maximum wavelength depends only on the temperature. The maximum wavelength is

$$\begin{aligned} \lambda_{\text{max}} &= \frac{b}{T} \\ &= \frac{2.8978 \times 10^{-3} \text{ m·K}}{2900\text{K}} \\ &= 9.9 \times 10^{-7} \text{ m} \end{aligned}$$

b is a constant whose value is determined by experiment.

The answer is (A).

144. A generator bus is a generic term for the bus in closest electrical contact to the generator. A load bus is any bus containing the system loads. A slack bus is the reference bus for the voltage angle. A voltage-controlled bus is any bus where the voltage is kept constant.

The answer is (C).

145. Non-dwelling receptacle outlet loads are calculated in accordance with NEC Sec. 220.14(I). A duplex or triplex receptacle load is calculated at 180 VA per unit. The quad or higher unit receptacle load is calculated at 90 VA per receptacle. Therefore, for a quad, the load is 360 VA. This combines for a total load of

$$(45)(180 \text{ VA}) + (5)(180 \text{ VA}) + (10)(360 \text{ VA}) = 12{,}600 \text{ VA}$$

Loads are calculated in Part II of Art. 220, then demand factors in Parts III, IV, and V are applied. The office space given is a non-dwelling unit. The demand factors for such receptacle loads are covered in NEC Sec. 220.44.

Two methods are allowed: one method is to combine the receptacle load with the lighting load, but the problem statement ruled this method out. Therefore, the appropriate demand factors are given in Table 220.44. For the first 10 kVA of the load, the demand factor is 100% (1.00). The remainder over 10 kVA uses a demand factor of 50% (0.50). Apply this to the total load.

$$\begin{aligned} \text{total calculated load}_{\text{VA}} &= (\text{total load})(\text{demand factors}) \\ &= (10{,}000 \text{ VA})(1.00) \\ &\quad + (2600 \text{ VA})(0.50) \\ &= 11{,}300 \text{ VA} \end{aligned}$$

According to Sec. 220.14, the loads are to be based on the nominal branch circuit voltages, which for calculation purposes are standardized in Sec. 220.5(A). In this case, 120 V is the given branch circuit voltage. The calculated load is

$$\begin{aligned} \text{total calculated load}_{\text{A}} &= \frac{\text{total calculated load}_{\text{VA}}}{\text{nominal voltage}} \\ &= \frac{11{,}300 \text{ VA}}{120 \text{ V}} \\ &= 94.17 \text{ A} \quad (94 \text{ A}) \end{aligned}$$

(Branch circuit and feeder calculations are described in NEC Art. 220.)

The answer is (A).

146. Option A is a model for a grounded wye load. Option B is a model for a grounded wye-delta transformer. Option C is a model for an ungrounded wye-delta transformer. Option D is a zero-sequence model for a delta-delta transformer.

The answer is (D).

147. Overload protection is also termed "overcurrent protection" and is covered in NEC Art. 240, which is the article used to start any overcurrent requirement search. In Sec. 240.3, the types of equipment requiring protection are listed along with the appropriate governing articles. From Table 240.3, the motors, motor circuits, and controllers requirements are given in Art. 430.

NEC Art. 430, Part III contains the overload requirements. The motor is greater than 1 hp, so the appropriate information is in Sec. 430.32(A)(1). The major settings are 115% and 125% of full-load current. The requirements are

requirements	overload setting (percentage of full-load nameplate current rating)
service factor not less than 1.15	125%
temperature rise not over 40°C	125%
all other motors	115%

The nameplate data for the given motor has a service factor of 1.00, which is less than 1.15, and a temperature rise of 60°C, which is greater than 40°C. Therefore, this motor is in the "all other motors" category. The overload setting is

$$\begin{aligned} &(\text{full-load current})(\text{Art. 430 overload setting}) \\ &\quad = (68 \text{ A})(1.15) \\ &\quad = 78.2 \text{ A} \end{aligned}$$

Since the article states that the overload setting can be no more than the overload setting percentage calculated, the appropriate setting is 70 A.

The answer is (B).

148. Option A describes the effect of an overvoltage condition. Option B refers to the effect of an undervoltage condition. Option C would result in an undervoltage condition. Option D states the result of an underfrequency condition. The effect on the turbine is due to stresses induced by deviation from synchronous speed.

The answer is (D).

149. Simple apparatus are defined in NEC Art. 100, Simple Apparatus, as having well-defined electrical parameters that do not generate more than 1.5 V, 100 mA, and 25 mW. LEDs are passive devices (see Informational Note 1 under simple apparatus in NEC Art. 100) and may exceed these parameters as long as the surface temperature does not exceed the ignition temperature of the hazardous material. The codes for the maximum surface temperature are in NEC Table 500.8(C). To determine the applicable code, the surface temperature must be calculated per NEC Sec. 504.10(D) as follows.

$$\begin{aligned} T_{\text{surface}} &= P_{\text{output}} R_{\text{thermal}} + T_{\text{ambient}} \\ &= (19.5 \text{ W})\left(9.6 \ \frac{^\circ\text{C}}{\text{W}}\right) + 40^\circ\text{C} \\ &= 227.2^\circ\text{C} \end{aligned}$$

Referring to NEC Table 500.8(C), the value of 227.2°C (230°C) means the applicable temperature code is T2C.

The answer is (A).

150. The term "reactor" indicates any device that introduces reactance of either type into a circuit. A reactor can be a capacitor or an inductor.

The answer is (C).

151. Find the per-unit impedance on the transmission line.

Per Unit System

$$Z_{base} = \frac{V_{base}^2}{S_{base}}$$

$$Z_{pu} = \frac{Z_{actual}}{Z_{base}} = \frac{Z_{actual}}{\frac{V_{base}^2}{S_{base}}}$$

The "power" is actually the apparent power, hence the use of the symbol S. Substitute the calculated and given information.

$$\begin{aligned} Z_{pu} &= \frac{Z_{actual}}{\frac{V_{base}^2}{S_{base}}} = \frac{30 + j100\ \Omega}{\left(\frac{(2\ \text{kV})^2}{10\ \text{kVA}}\right)\left(1000\ \frac{\Omega}{\text{k}\Omega}\right)} \\ &= \frac{30 + j100\ \Omega}{400\ \Omega} \\ &= 0.075 + j0.250\ \text{pu} \end{aligned}$$

The answer is (C).

152. The short-circuit current downstream of the transformer is limited by the 2 MVA rating of the transformer. The impedance that the current must pass through includes the generator's impedance and the transformer's impedance. Because the X/R ratios are identical, the ratio of the powers provides the per-unit value of the generator's impedance, in this case, 2 MVA/20 MVA or 0.1 pu. Find the short-circuit current.

Fault Current Analysis

$$S_{sc} = \sqrt{3}\, V_{nom} I_{sc}$$

$$\begin{aligned} I_{sc} &= \frac{I_{sc}}{\sqrt{3}\, V_{base}} = \frac{\frac{S_{base}}{Z_{pu}}}{\sqrt{3}\, V_{base}} \\ &= \frac{\frac{2 \times 10^6\ \text{VA}}{0.05\ \text{pu} + 0.1\ \text{pu}}}{\sqrt{3}(460\ \text{V})} \\ &= 1.67 \times 10^4\ \text{A} \quad (20\ \text{kA}) \end{aligned}$$

The answer is (B).

153. The circuit shown is a connected delta. Therefore,

3-Phase Circuits

$$V_{line} = V_{phase} = 120\ \text{V}$$

The answer is (D).

154. To design a machine of the same power with half the rated speed, the number of conductors must be increased by twice the original value, or the flux can be doubled. Either choice increases the size of the machine. In practice, both methods are used. As a result, for a given power output, a lower speed machine will always be larger than a higher speed machine. Therefore, options A, B, and C are true. Option D is false.

The answer is (D).

155. Ohm's law can be used to determine the current through the resistor. Find the voltage on the secondary side.

Single-Phase Transformer Equivalent Circuits

$$\frac{E_1}{E_2} = \frac{N_1}{N_2} = \frac{V_1}{V_2}$$

$$\frac{V_p}{V_s} = \frac{N_p}{N_s}$$

$$\begin{aligned} V_s &= V_p\left(\frac{N_s}{N_p}\right) \\ &= (120\ \text{V})\left(\frac{1}{20}\right) \\ &= 6\ \text{V} \end{aligned}$$

Apply Ohm's law to the secondary.

$$V_{\text{load}} = I_{\text{load}} R_{\text{load}}$$
$$I_{\text{load}} = \frac{V_{\text{load}}}{R_{\text{load}}} = \frac{6 \text{ V}}{2 \ \Omega} = 3 \text{ A}$$

The answer is (A).

156. The terminal voltage is

$$\begin{aligned} V_{\text{terminal}} &= E_{\text{source}} - I_{\text{load}} R_{\text{internal}} \\ &= 12 \text{ V} - (4 \text{ A})(0.1 \ \Omega) \\ &= 11.6 \text{ V} \quad (12 \text{ V}) \end{aligned}$$

The answer is (D).

157. The rated power is [Conversion Factor]

$$\begin{aligned} P_{\text{rated}} &= (10 \text{ hp})\left(745.7 \ \frac{\text{W}}{\text{hp}}\right) \\ &= 7457 \text{ W} \end{aligned}$$

The input power depends on the efficiency and is determined from the following.

Energy Management

$$\%\text{efficiency} = \left(\frac{P_{\text{out}}}{P_{\text{in}}}\right) \times 100$$
$$\begin{aligned} P_{\text{in}} &= \frac{P_{\text{out}}}{\eta} = \frac{P_{\text{rated}}}{\eta} \\ &= \frac{7457 \text{ W}}{0.9} \\ &= 8285.6 \text{ W} \quad (8300 \text{ W}) \end{aligned}$$

The answer is (D).

158. The synchronous motor is 250 VA at a power factor of 0.8 leading.

$$\begin{aligned} \mathbf{S}_{\text{sync}} &= 250 \text{ VA}\angle \arccos 0.8 \\ &= 250 \text{ VA}\angle -36.87^\circ \end{aligned}$$

The induction motor is 75 VA at a power factor of 0.7 lagging.

$$\begin{aligned} \mathbf{S}_{\text{induction}} &= 75 \text{ VA}\angle \arccos 0.7 \\ &= 75 \text{ VA}\angle 45.57^\circ \end{aligned}$$

Add the two values to find the system power factor.

$$\begin{aligned} \mathbf{S}_{\text{system}} &= \mathbf{S}_{\text{sync}} + \mathbf{S}_{\text{induction}} \\ &= 250 \text{ VA}\angle -36.87^\circ + 75 \text{ VA}\angle 45.57^\circ \\ &= 270 \text{ VA}\angle -20.90^\circ \end{aligned}$$

The power factor of the system is

$$\begin{aligned} \text{pf} &= \cos(-20.90^\circ) \\ &= 0.93 \text{ leading} \quad (0.90 \text{ leading}) \end{aligned}$$

The answer is (D).

159. In a series DC motor, the armature current, unimpeded by counter-emf, flows through the series winding during startup, resulting in the most torque for the motors listed. (The wound rotor is an AC induction motor type.)

The answer is (B).

160. The signal in option A is for a half-wave rectifier. The signal in option B is for a full-wave bridge rectifier. The signal in option C is a base clipper and matches the signal shown. Option D is a peak clipper.

The answer is (C).